THE CONCEPT OF CREATIVITY IN SCIENCE AND ART

MARTINUS NIJHOFF PHILOSOPHY LIBRARY

VOLUME 6

Other volumes in this series:

1. Lamb, D.: Hegel — From foundation to system. 1980. ISBN 90-247-2359-0
2. Bulhof, I.N.: Wilhelm Dilthey — a hermeneutic approach to the study of history and culture. 1980. ISBN 90-247-2360-4
3. van der Dussen, W.J.: History as a science — the philosophy of R.G. Collingwood. ISBN 90-247-2453-8
4. Chatterjee, M.: The language of philosophy. ISBN 90-247-2372-8
5. Kluge, E.-H.W.: The metaphysics of Gottlob Frege — an essay in ontological reconstruction. 1980. ISBN 90-247-2422-8

ISBN Martinus Nijhoff Philosophy Library series 90-247-2344-2

AMERICAN UNIVERSITY PUBLICATIONS IN PHILOSOPHY

III

THE CONCEPT OF CREATIVITY IN SCIENCE AND ART

edited by Denis Dutton and Michael Krausz

editors of the series Harold A. Durfee and David F.T. Rodier

Other volumes in this series:

I. The Faculty in Philosophy at the American University: Explanation — new directions in philosophy. 1973. ISBN 90-247-1517-2
II. Durfee, H.A.: Analytical philosophy and phenomenology. 1976. ISBN 90-247-1880-5

Published by Martinus Nijhoff Publishers

THE CONCEPT OF CREATIVITY IN SCIENCE AND ART

edited by

DENIS DUTTON and MICHAEL KRAUSZ

1981

MARTINUS NIJHOFF PUBLISHERS
THE HAGUE / BOSTON / LONDON

Distributors:

for the United States and Canada

Kluwer Boston, Inc.
190 Old Derby Street
Hingham, MA 02043
USA

for all other countries

Kluwer Academic Publishers Group
Distribution Center
P.O. Box 322
3300 AH Dordrecht
The Netherlands

Library of Congress Cataloging in Publication Data
Main entry under title:

The Concept of creativity in science and art.

(Martinus Nijhoff philosophy library; v. 6)
(American University publications in philosophy; 3)
Includes—index.
Contents: The three domains of creativity /
A. Koestler —— Creativity in science / R. Harré ——
Evey horse has a mouth / F.E. Sparshott —— [etc.]
1. Creation (Literary, artistic, etc.) —— Addresses,
essays, lectures. 2. Creative ability in science ——
Addresses, essays, lectures. 3. Creative thinking
—— Addresses, essays, lectures. I. Dutton, Denis.
II. Krausz, Michael. III. Series. IV. Series:

American University publications in philosophy; 3.
BH301.C84C66   128'.3   81-4001
ISBN 90-247-2418-X   AACR2
ISBN 90-247-2344-2 (Martinus Nijhoff Philosophy Library series)

PRINTED IN THE NETHERLANDS

TABLE OF CONTENTS

List of editors and contributors   vii

Series editors' preface   ix

Editors' preface   xi

The three domains of creativity
A. KOESTLER   1

Creativity in science
R. HARRÉ   19

Every horse has a mouth: a personal poetics
F.E. SPARSHOTT   47

Criteria of creativity
C.R. HAUSMAN   75

The creative imagination
M. POLANYI   91

The rationality of creativity
I.C. JARVIE   109

Creative product and creative process in science and art
L. BRISKMAN   129

Creativity as learning process
C.A. VAN PEURSEN   157

Creating and becoming
M. KRAUSZ   187

On the dialectical phenomenology of creativity
A. HOFSTADTER   201

Name index   209

LIST OF EDITORS AND CONTRIBUTORS

Editors:

Denis Dutton, Department of Philosophy, University of Michigan –
Dearborn, Dearborn, Michigan 48128, USA.

Michael Krausz, Department of Philosophy, Bryn Mawr College,
Bryn Mawr, Pennsylvania 19010, USA.

Contributors:

Larry Briskman, Department of Philosophy, University of
Edinburgh, Edinburgh, Scotland EH8 9JX, Great Britain.
Rom Harré, Linacre College, Oxford University, Oxford OX1 3PN,
England.
Carl Hausman, Department of Philosophy, Pennsylvania State
University, University Park, Pennsylvania 16802, USA.
Albert Hofstadter, Department of Philosophy, University of
California, Santa Cruz, California 95060, USA.
I.C. Jarvie, Department of Philosophy, York University, Downsview,
Ontario M3J 1P3, Canada.
Arthur Koestler, London, England.
C.A. van Peursen, Rijksuniversiteit Leiden, Centrale Interfaculteit,
Filosofisch Instituut, Witte Singel 71, Leiden, The Netherlands.
Michael Polanyi, late of the University of Chicago.
F.E. Sparshott, Department of Philosophy, Victoria College,
University of Toronto, Toronto, Ontario M5S 1A1, Canada.

SERIES EDITORS' PREFACE

This third volume of American University Publications in Philosophy continues the tradition of presenting books in the series shaping current frontiers and new directions in philosophical reflection. In a period emerging from the neglect of creativity by positivism, Professors Dutton and Krausz and their eminent colleagues included in the collection challenge modern philosophy to explore the concept of creativity in both scientific inquiry and artistic production. In view of the fact that Professor Krausz served at one time as Visiting Professor of Philosophy at The American University we are especially pleased to include this volume in the series.

HAROLD A. DURFEE, for the editors of
American University Publications in Philosophy

EDITORS' PREFACE

While the literature on the psychology of creativity is substantial, surprisingly little attention has been paid to the subject by philosophers in recent years. This fact is no doubt owed in part to the legacy of positivism, whose tenets have included a sharp distinction between what Hans Reichenbach called the context of discovery and the context of justification. Philosophy in this view must address itself to the logic of justifying hypotheses; little of philosophical importance can be said about the more creative business of discovering them. That, positivism has held, is no more than a merely psychological question: since there is no logic of discovery or creation, there can be no philosophical reconstruction of it.

Though the field of aesthetics has naturally addressed itself to the question of creativity with greater frequency than empiricist philosophy of science, even there attention in the last two decades has been largely concentrated on such issues as the logic of critical interpretation or the ontology of works of art. The question of artistic creativity has been once again left to the hands of psychologists.

We present this collection of articles in the belief that there are questions of a significantly philosophical sort to be asked about the concept of creativity, questions which overlap the fields of both art and science.

We wish to acknowledge the permission granted by the McGraw-Hill Book Company to reprint the article by Arthur Koestler, *Chemical and Engineering News* to include Michael Polanyi's contribution, and *Philosophy and Phenomenological Research* to present Carl Hausman's paper. The other articles were written especially for this volume, though F.E. Sparshott's contribution has been published in *Philosophy and Literature* and Larrv Briskman's has appeared in *Inquiry*. Both appear with the kind permission of the publishers. We appreciate the capable assistance of Daniel Marowski and David Sosnowski in the preparation of the manuscript.

DENIS DUTTON
University of Michigan – Dearborn
MICHAEL KRAUSZ
Bryn Mawr College

THE THREE DOMAINS OF CREATIVITY*

ARTHUR KOESTLER

This paper attempts to give a condensed outline of a theory I have
set out in detail in my book on creativity and to carry that theory
to the next step.[1] The proposition I shall submit is, in a nutshell,
that the conscious and unconscious processes which enter into all
three forms of creative activity have a basic pattern in common.
And when I speak of *three* forms of creativity, I mean the domains
of artistic originality, scientific discovery, and comic inspiration.
I believe that all creative activity falls into one or another of these
three categories or, more frequently, into some combination of
them. If you speak, for instance, of cooking as "creative," you
automatically imply that cooking is either an art or a science or
both.

As a first step toward describing that pattern, let us try something simple like this: The creative act consists in combining
previously unrelated structures in such a way that you get more
out of the emergent whole than you have put in. This sounds like
making a *perpetuum mobile,* and in a sense it is, because mental
evolution, like biological evolution, seems to contradict the second
law of thermodynamics, which contends that the universe is
running down as if afflicted by mental fatigue. But we will not go
into this; instead, let me illustrate by a few schoolbook examples
what I mean by combining two previously unrelated structures.

I. Association and Bisociation

The motions of the tides have been known to man since time
immemorial. So have the motions of the moon. But the idea to
relate the two, the idea that the tides were due to the attraction
of the moon, occurred, as far as we know, for the first time to a
German astronomer in the seventeenth century, and when Galileo

D. Dutton and M. Krausz, eds., The Concept of Creativity in Science and Art, pp. 1-17.

read about it, he laughed it off as an occult fancy.[2] Moral: The more familiar each of the previously unrelated structures are, the more striking the new synthesis and the more obvious it seems in the driver's mirror of hindsight.

The history of science is a history of marriages between ideas which were previously strangers to each other, and frequently considered incompatible. Lodestones — magnets — were known in antiquity as some curiosity of nature. In the Middle Ages they were used for two purposes: as navigators' compasses and as a means to attract an estranged wife back to her husband. Equally well known were the curious properties of amber, which, when rubbed, acquires the virtue of attracting flimsy objects. The Greek word for amber is *elektron*, but the Greeks were not much interested in electricity, nor were the Middle Ages. For nearly two thousand years, electricity and magnetism were considered separate phenomena, in no way related to each other. In 1820, Hans Christian Oersted discovered that an electric current flowing through a wire deflected a compass needle which happened to be lying on his table. At that moment the two contexts began to fuse into one — electromagnetism — creating a kind of chain reaction which is still continuing and gaining in momentum; forever amber.

From Pythagoras, who combined arithmetic and geometry, to Einstein, who unified energy and matter in a single sinister equation, the pattern is always the same. The Latin word *cogito* comes from *coagitare*, "to shake together." The creative act does not create something out of nothing, like the God of the Old Testament; it combines, reshuffles, and relates already existing but hitherto separate ideas, facts, frames of perception, associative contexts. This act of cross-fertilization — or self-fertilization within a single brain — seems to be the essence of creativity. I have proposed for it the term *bisociation*. It is not a pretty word, but it helps us to make a distinction between the sudden leap of the creative act and the more normal, more pedestrian, associative routines of thinking.

The difference between the two could be described as follows. Orderly thinking (as distinct from daydreaming) is always controlled by certain rules of the game. In the psychological laboratory, the experimenter lays down the rule: "Name opposites." Then he says "dark," and the subject promptly says, "light." But if the rule is "synonyms," then the subject will associate "dark" with "black" or "night" or "shadow." To talk of stimuli in a vacuum is meaningless; what response a stimulus will evoke depends on the game we are playing at the time.

But we do not live in laboratories where the rules of the game are laid down by explicit orders; in normal life, the rules control our thinking unconsciously — and there's the rub. When talking, the laws of grammar and syntax function below the level of awareness, in the gaps between the words. So do certain simple rules of common or garden-variety logic and of courtesy and convention, and also the complex and specialized rules which we call "frames of reference" or "universes of discourse" or "thinking in terms of" this or that — of physiological explanations or ethical value judgments. All thinking is playing a game according to fixed rules and more or less flexible strategies. The game of chess allows you a vast number of strategic choices among the moves permitted by the rules, but there is a limit to them. There are hopeless situations in chess when the most subtle strategies will not save you — short of offering your opponent a jumbo-sized Martini. Now in fact there is no rule in chess preventing you from offering your opponent a Martini. But making a person drunk while remaining sober oneself is a different sort of game with a different context. Combining the two games is a bisociation. In other words, associative routine means thinking according to a given set of rules, on a single plane, as it were. The bisociative act means combining two different sets of rules, to live on several planes at once.

II. Three Kinds of Reactions

I do not mean to belittle the value of law-abiding routines. They lend coherence and stability to behavior and structured order to thought. But they have their obvious limitations. For one thing, every game tends to become monotonous after awhile and fails to satisfy the artist's craving for self-expression and the scientist's search for explanations. In the second place, the world moves on, and new problems arise which cannot be solved within the conventional frames of reference by applying to them the accepted rules of the game. Then a crisis occurs: The position on the scientist's checkerboard is blocked;the artist's vision is blurred;the search is on, the fumbling and groping for that happy combination of ideas — of lodestone and amber — which will lead to the new synthesis.

The Aha Reaction. Gestalt psychologists have coined a word for that moment of truth, the flash of illumination, when bits of the puzzle suddenly click into place. They call it the *Aha* experience. One may regard it as a synonym for the "Eureka!" cry. Imagine it written on a blackboard, thus:

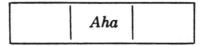

We shall see in a moment the reason for this display. There is an empty panel on each side — for the *Aha* response represents only *one* type of reaction after the bisociative click. There are others. Let me tell my favorite anecdote:

> Λ nobleman at the court of Louis XV had unexpectedly returned from a journey and, on entering his wife's boudoir, found her in the arms of a bishop. After a short hesitation, the nobleman walked calmly to the window and went through the motions of blessing the people in the street.
> "What are you doing?" cried the anguished wife.
> "Monseigneur is performing my functions," replied the nobleman, "so I am performing his."

Well, some readers will be kind enough to laugh; let us call this the *Haha* reaction:[3]

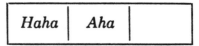

The Haha Reaction. Now let us inquire into the difference between the *Haha* and the *Aha* reactions. Why do we laugh? Let me try to analyze first the intellectual and then the emotional aspect of this odd reaction. The nobleman's behavior is both unexpected and perfectly logical — but of a logic not usually applied to this type of situation. It is the logic of the division of labor, where the rule of the game is the *quid pro quo,* the give-and-take. But we expected, of course, that his reactions would be governed by a quite different logic or rule of the game. It is the interaction between these two mutually exclusive associative contexts which produces the comic effect. It compels us to perceive the situation at the same time in two self-consistent but habitually incompatible frames of reference; it makes us function on two wavelengths simultaneously, as it were. While this unusual condition lasts, the event is not, as is normally the case, perceived in a single frame of reference but is bisociated with two.

But the unusual condition does not last for long. The act of discovery leads to a lasting synthesis, a *fusion* of the two previously unrelated frames of reference; in the comic bisociation you have a *collision* between incompatible frames which for a brief moment cross each other's path. But whether the frames are compatible or not, whether they will collide or merge, depends on subjective factors, on the attitudes of the audience — for, after all, the colliding or merging takes place in the audience's heads. The history of science abounds with examples of discoveries greeted with howls of laughter because they seemed to be a marriage of incompatibles — until the marriage bore fruit and the alleged incompatibility of the partners turned out to derive from prejudice. The humorist, on the other hand, deliberately chooses discordant codes of behavior or universes of discourse to expose their hidden incongruities in the resulting clash. Comic discovery is paradox stated — scientific discovery is paradox resolved.

Let me return for a moment to our poor nobleman blessing the crowd through the window. His gesture was a truly original inspiration. If he had followed the conventional rules of the game, he would have had to beat up or kill the bishop. But at the court of Louis XV, assassinating a monseigneur would have been considered, if not exactly a crime, still in very bad taste. It simply could not be done; the chessboard was blocked. To solve the problem, that is, to save his face and at the same time to humiliate his opponent, the nobleman had to bring into the situation a second frame of reference, governed by different rules of the game, and combine it with the first. All original comic invention is a creative act, a malicious discovery.

The Emotional Dynamics of Laughter. The emphasis is on malicious, and this brings us from the *logic* of humor to the *emotional factor* in the *Haha* reaction. When the expert humorist tells an anecdote, he creates a certain tension which mounts as the narrative progresses. But it never reaches its expected climax. The punch line acts like a guillotine which cuts across the logical development of the situation; it debunks our dramatic expectations, and the tension becomes redundant and is exploded in laughter. To put it differently, laughter disposes of the overflow of emotion which has become pointless, is denied by reason, and has to be somehow worked off along physiological channels of least resistance.[4]

I shall not bore you with physiological explanations because if you look at the coarse and brutal merriment in a tavern scene by Hogarth or Rawlinson, you realize at once that the revelers are working off their surplus of adrenalin by contractions of the face muscles, slapping of thighs, and explosive exhalations of breath from the half-closed glottis. The emotions worked off in laughter are aggression, sexual gloating, conscious or unconscious sadism — all operating through the sympathicoadrenal system. On the other hand, when you look at a clever *New Yorker* cartoon, Homeric laughter yields to an amused and rarefied smile; the ample flow of adrenalin has been distilled into a grain of Attic salt. Think, for instance, of that classic definition: "What is a sadist?" "A person who is kind to a masochist."

The word "witticism" is derived from "wit" in its original sense of "ingenuity." The clown is brother to the sage; their domains are continuous, without a sharp dividing line. As we move from the coarse toward the higher forms of humor, the joke shades into epigram and riddle, the comic simile into the discovery of hidden analogies; and the emotions involved show a similar transition. The emotive voltage discharged in coarse laughter is aggression robbed of its purpose; the tension discharged in the *Aha* reaction is derived from an intellectual challenge. It snaps at the moment when the penny drops — when we have solved the riddle hidden in the *New Yorker* cartoon, in a brainteaser, or in a scientific problem.

The Ah Reaction. Let me repeat, the two domains of humor and discovery form a continuum. As we travel across it, from left to center, so to speak, the emotional climate gradually changes from the malice of the jester to the detached objectivity of the sage. If we now continue the journey in the same direction, we find equally gradual transitions into a third domain, that of the artist. The artist hints rather than states, and he poses riddles. So we get a symmetrically reversed transition toward the other end of the spectrum, from highly intellectualized art forms toward the more sensual and emotive, ending in the thought-free beatitude of the oceanic feeling — the cloud of unknowing.

But how does one define the emotional climate of art? How does one classify the emotions which give rise to the experience of beauty? If you leaf through textbooks of experimental psychology, you will not find much mention of it. When behaviorists use the word "emotion," they nearly always refer to hunger, sex, rage, and fear and to the related effects of the release of adrenalin. They have no explanations to offer for the curious reaction one experiences when listening to Mozart of looking at the ocean or reading for the first time John Donne's *Holy Sonnets*. Nor will you find in the textbooks a description of the physiological processes accompanying the reaction: the moistening of the eyes, perhaps a quiet overflow of the lachrymal glands, the catching of one's breath, followed by a kind of rapt tranquillity, the draining of all tensions. Let us call this the *Ah* reaction and thus complete our trinity.

Haha!	*Aha*	*Ah. . . .*

Laughter and weeping, the Greek masks of comedy and tragedy, mark the two extremes of a continuous spectrum; both are overflow reflexes, but they are in every respect physiological opposites. Laughter is mediated by the sympathicoadrenal branch of the autonomous nervous system, weeping by the parasympathetic branch. The first tends to galvanize the body into action; the second tends toward passivity and catharsis. Watch how you breathe when you laugh: long, deep intakes of air, followed by bursts of exhalatory puffs — "Ha, ha, ha." In weeping, you do the opposite: short, gasping inspirations — sobs — are followed by long, sighing expirations — "a-a-h, aah."[5]

III. Self-assertion and Self-transcendence

In keeping with this, the emotions which overflow in the *Ah* reaction are the direct opposites of those exploded in laughter. The latter belong to the familiar adrenergic hunger-rage-fear category; let us call them the *aggressive-defensive* or self-assertive emotions. Their opposites we might call the *self-transcending* or participatory or integrative emotions. They are epitomized in what Freud called the "oceanic feeling": When you listen to a Bach toccata thundering through the cathedral, you experience that expansion and depersonalization of awareness in which the self seems to dissolve like a grain of salt in a lot of water.

This class of emotions shows a wide range of variety. They may be joyous or sad, tragic or lyrical; but they have a common denominator: the feeling of participation in an experience which transcends the boundaries of the self. That higher entity, of which the self feels a part, to which it surrenders its identity, may be nature, God, the anima mundi, the magic of forms, or the ocean of sound.

The self-assertive emotions are expressed in bodily actions; the self-transcending emotions operate through the passive processes of empathy, rapport, projection, and identification. In laughter, tension is suddenly exploded, emotion debunked; in weeping, it is drained away in a gradual process which does not break the continuity of mood. The self-transcending emotions do not tend toward action but toward quiescence and catharsis. Respiration and pulse rate are slowed down; "entrancement" is a step toward the trancelike states induced by contemplative techniques or drugs. The self-transcending emotions cannot be consummated by any specific, voluntary action. You cannot take the mountain panorama home with you; you cannot merge with the infinite by any exertion of the body. To be "overwhelmed" by awe and wonder, "enraptured" by a smile, "entranced" by beauty — each of these verbs expresses a passive surrender. The surplus of emotion cannot be worked off in action; it can be consummated only in internal, visceral, and glandular processes.[6]

The participatory or self-transcending tendencies, these stepchildren of psychology, are as powerful and deeply rooted in man's nature as his self-assertive drives. Freud and Piaget, among others, have emphasized the fact that the very young child does not differentiate between ego and environment. The nourishing breast appears to it as a more intimate possession than the toes of its own body. It is aware of events but not of itself as a separate entity. It lives in a state of mental symbiosis with the outer world, a continuation of the biological symbiosis in the womb. The universe is focused in the self, and the self *is* the universe — a condition which Piaget called "protoplasmic consciousness." It may be likened to a liquid, fluid universe, traversed by dynamic currents, the rise and fall of physiological needs causing minor storms which come and go without leaving solid traces. Gradually the floods recede, and the first islands of objective reality emerge; the contours grow firmer and sharper; the islands grow into continents, the dry territories of reality are mapped out; but side by side with it, the liquid world coexists, surrounding it, inter-penetrating it by canals and inland lakes, the vestigial relics of the erstwhile symbiotic communion. Here, then, we have the origin of that oceanic feeling which the artist and the mystic strive to recapture on a higher level of development, at a higher turn of the spiral.

IV. Art and Self-transcendence

Children and primitives are apt to confuse dream and reality; they not only believe in miracles but also believe themselves capable of performing them. When the medicine man disguises himself as the rain god, he produces rain. Drawing a picture of a slain bison assures a successful hunt. This is the ancient unitary source out of which the ritual dance and song, the mystery plays of the Achaeans, and the calendars of the Babylonian priest-astronomers were derived. The shadows in Plato's cave are symbols of man's loneliness; the paintings in the Altamira caves are symbols of his magic powers.

We have traveled a long way from Altamira and Lascaux, but the artist's inspirations and the scientist's intuitions are still fed by that same unitary source — though by now we should rather call it an underground river. Wishes do not displace mountains, but in our dreams they still do. Symbiotic consciousness is never completely defeated but merely relegated underground to those unconscious levels in the mental hierarchy where the boundaries of the ego are still fluid and blurred — as blurred as the distinction between the actor and the hero whom he impersonates and with whom the spectator identifies. The actor on the stage is himself and somebody else at the same time — he is both the dancer and the rain god.

Dramatic illusion is the coexistence in the spectator's mind of two universes which are logically incompatible; his awareness, suspended between the two planes, exemplifies the bisociative process in its most striking form. All the more striking because he produces physical symptoms — palpitations, sweating, or tears — in response to the perils of a Desdemona whom he *knows* to exist merely as a shadow on the TV screen or as dry printer's ink in the pages of a book. Yet let Othello but get the hiccups, and instead of coexistence between the two planes juxtaposed in the spectator's mind, you get collision between them. Comic impersonation produces the *Haha* reaction because the parodist arouses aggression and malice; dramatic stagecraft achieves the suspension of disbelief, the coexistence of incompatible planes, because it induces the spectator to identify. It excites the self-transcending and inhibits or neutralizes the self-assertive emotions. Even when fear and anger are aroused in the spectator, these are vicarious emotions, derived from his identification with the hero, which in itself is a self-transcending act. Vicarious emotions aroused in this manner carry a dominant element of sympathy, which facilitates catharsis in conformity with the Aristotelian definition — "through incidents arousing horror and pity to accomplish the purgation of such emotions." Art is a school of self-transcendence.

We thus arrive at a further generalization: *The* Haha *reaction signals the collision of bisociated contexts; the* Aha *reaction signals their fusion; and the* Ah *reaction signals their juxtaposition.*

This difference is reflected in the quasicumulative progression of science through a series of successive mergers, compared with the quasi-timeless character of art in its continuous restatement of basic patterns of experience in changing idioms. I said "quasi" because it can be shown that this, too, is a matter of degrees, because the progress of science is not cumulative in the strict sense.

It is moving in a dizzy, zigzag course rather than in a straight line.[7] On the other hand, the development of a given art form over a period of time often displays a cumulative progression.[8] I shall return to this in a moment, but first let me briefly mention a few more types of the combinatorial activities which enter into the fabric of art.

V. Bisociative Structures in Art

When we listen to poetry, two frames of reference interact in our minds: one governed by meaning, the other by rhythmic patterns of sound. Moreover, the two frames operate on two different levels of awareness: the first in broad daylight, the other much deeper down. The rhythmic beat of poetry is designed, in the words of Yeats, "to lull the mind into a waking trance." Rhythmic pulsation is a fundamental characteristic of life; our ready responses to it arise from the depths of the nervous system, from those archaic strata which reverberate to the shaman's drum and which make us particularly receptive to, and suggestible by, messages which arrive in a rhythmic pattern or are accompanied by such a pattern.

The rhyme has equally ancient roots. It repeats the last syllable of a line. Now the repetition of syllables is a conspicuous phenomenon at the very origins of language. The young child is addicted to babbling, "obble-gobble," "humpty-dumpty," and so on. In primitive languages, words like "kala-kala" or "moku-moku" abound. Closely related to it is association by pure sound. The rhyme is in fact nothing but a glorified pun — two strings of ideas tied together in a phonetic knot. Its ancient origins are revealed in the punning mania of children and in certain forms of mental disorder and in the frequent recurrence of puns in dreams. "What could be moister than the tears of an oyster?" The statement that the oyster is a wet creature and that therefore its tears must be particularly wet would not make much of an impression, but when meaning is bisociated with sound, there is magic. This is what I meant when I said that routine thinking involves a single matrix, whereas creative thinking always involves more than one plane. Needless to say, it is difficult to identify with an oyster, so the reaction will be *Haha,* not *Ah.*

Thus rhythm and meter, rhyme and euphony, are not artificial ornaments of language but combinations of contemporary, sophisticated frames of reference with archaic and emotionally more powerful games of the mind. In other words, creative activity always implies a *temporary regression* to these archaic levels, while a simultaneous process goes on in parallel on the highest, most articulate and critical level: the poet is like a skin diver with a breathing tube.

The same applies, of course, to poetic imagery. Visual thinking is an earlier form of mental activity than thinking in verbal concepts; we dream mostly in pictures and visual symbols. It has been said that scientific discovery consists in seeing an analogy where nobody has seen one before. When in the *Song of Songs,* Solomon compared the Shulamite's neck to a tower of ivory, he saw an analogy which nobody had seen before; when Harvey compared the heart of a fish to a mechanical pump, he did the same; and when the caricaturist draws a nose like a cucumber, he again does just that. In fact, all combinatorial, bisociative patterns are trivalent — they can enter the service of humor, discovery, or art, as the case may be.

Let me give you another example of this trivalence. Man has always looked at nature by superimposing a second frame on the retinal image — mythological, anthropomorphic, scientific frames. The artist sees in terms of his medium — stone, clay, charcoal, pigment — and in terms of his preferential emphasis on contours or surfaces, stability or motion, curves or cubes. So, of course, does the caricaturist, only his motives are different. And so does the scientist. A geographical map has the same relation to a landscape that a character sketch has to a face. Every diagram or model, every schematic or symbolic representation of physical or mental processes is an unemotional caricature of reality — at least unemotional in the sense that the bias is not of an obvious kind, although some models of the human mind as a conditioned-reflex automaton seem to be crude caricatures inspired by unconscious bias.

In the language of behaviorist psychology, we would have to say that Cézanne, glancing at a landscape, receives a stimulus, to which he responds by putting a dab of paint on the canvas, and that is all there is to it. But in fact the two activities take place on two different planes. The stimulus comes from one environment, the distant landscape. The response acts on a different environment, a square surface of 10 by 15 inches. The two environments obey two different sets of laws. An isolated brushstroke does not represent an isolated detail in the landscape. There are no point-to-point correspondences between the two planes; each obeys a different rule of the game. The artist's vision is bifocal, just as the poet's voice is bivocal, as he bisociates sound and meaning.

VI. Extraconscious Factors in Discovery

Let me return for a moment to science. I said at the beginning of this paper that the essence of discovery is the coagitation, the shaking together, of already existing frames of reference or areas of knowledge. Now we arrive at the crucial question: Just how does the creative mind hit upon that happy combination of ideas which nobody had thought of combining before?

Artists are inclined to believe that scientists reason in strictly rational, precise verbal terms. They do, of course, nothing of the sort. In 1945, a famous inquiry was organized by Jacques Hadamard among eminent mathematicians in America to find out their working methods.[9] The results showed that all of them, with only two exceptions, thought neither in verbal terms nor in algebraic symbols but relied on visual imagery of a vague, hazy kind. Einstein was among those who answered the questionnaire; he wrote: "The words of the language as they are written or spoken do not seem to play any role in my mechanism of thought, which relies on more or less clear images of a visual and some of a muscular type. It seems to be that what you call full consciousness is a limiting case which can never be fully accomplished because consciousness is a narrow thing."

Einstein's statement is typical. On the testimony of those original thinkers who have taken the trouble to record their methods of work, *not only verbal thinking but conscious thinking in general plays only a subordinate part in the brief, decisive phase of the creative act itself.* Their virtually unanimous emphasis on spontaneous intuitions and hunches of unconscious origin, which they are at a loss to explain, suggests that the role of strictly rational and verbal processes in scientific discovery has been vastly over-estimated since the age of enlightenment. There are always large chunks of irrationality embedded in the creative process, not only in art (where we are ready to accept it) but in the exact sciences as well.

The scientist who, facing his blocked problem, regresses from precise verbal thinking to vague visual imagery seems to follow Woodworth's advice: "Often we have to get away from speech in order to think clearly." Words crystallize thoughts, but a crystal is no longer a liquid. Language can act as a screen between the thinker and reality. Creativity often starts where language ends, that is, by regressing to preverbal levels, to more fluid and uncommitted forms of mental activity.

Now I do not mean, of course, that there is a little Socratic demon housed in the scientist's or artist's skull who does his homework for him; nor should one confuse unconscious menta-tion with Freud's primary process. The primary process is defined by him as devoid of logic, governed by the pleasure principle, apt to confuse perception and hallucination, and accompanied by massive discharges of affect. It seems that between this very primary process and the so-called secondary process governed by the reality principle, we must interpolate a whole hierarchy of cognitive structures, which are not simply mixtures of primary and secondary, but are autonomous systems in their own right, each governed by a distinct set of rules. The paranoid delusion, the dream, the daydream, free association, the mentalities of children of various ages and of primitives at various stages should not be lumped together, for each has its own logic or rules of the game. But while clearly different in many respects, all these forms of mentation have certain features in common, since they are ontogenetically, and perhaps phylogenetically, older than those of the civilized adult. They are less rigid, more tolerant, and more ready to combine seemingly incompatible ideas and to perceive hidden analogies between cabbages and kings. One might call them "games of the underground," because if not kept under restraint, they would play havoc with the routines of disciplined thinking. But under exceptional conditions, when disciplined thinking is at the end of its tether, a temporary indulgence in these underground games may suddenly produce a solution — some farfetched, reckless combination which would be beyond the reach of, or seem to be unacceptable to, the sober, rational mind. The place for the rendezvous of ideas is underground.

Illumination and Catharsis. What I have been trying to suggest is that the common pattern underlying scientific discovery and artistic inspiration is a temporary regression, culminating in the bisociative act, i.e., the bringing together of previously separate frames of perception or universes of discourse. I suppose that is what Ernst Kris meant by his frequently quoted but somewhat cryptic remarks about regression in the service of the ego.[10] The boundaries between science and art, between the *Ah* reaction and the *Aha* reaction, are fluid, whether we consider architecture or cooking or psychiatry or the writing of history. There is nowhere a sharp break where witticism changes into wit or where science stops and art begins. Science, the hoary cliché goes, aims at truth, art at beauty. But the criteria of truth, such as verification by experiment, are not as hard and clean as we tend to believe, for the same experimental data can often be interpreted in more than one way. That is why the history of science echoes with as many bitter and venomous controversies as the history of literary criticism. Moreover, the verification of a discovery comes after the act; the creative act itself is for the scientist, as it is for the artist, a leap into the dark, where both are equally dependent on their fallible intuitions. The greatest mathematicians and physicists have confessed that, at those decisive moments when taking the plunge, they were guided not by logic but by a sense of beauty which they were unable to define. Vice versa, painters and sculptors, not to mention architects, have always been guided and often obsessed by scientific or pseudo-scientific theories and criteria of truth: the golden section, the laws of perspective, Dürer's and Leonardo's laws of proportion representing the human body, Cézanne's doctrine that everything in nature is modeled on the cylinder and cone, Braque's alternative theory that cubes should be substituted for spheres, le Courbusier's modulator theory, Buckminster Fuller's geodesic domes. The same goes, of course, for literature, from the formal laws imposed on Greek tragedy to the various recent and contemporary schools —

romanticism, classicism, naturalism, symbolism, stream of consciousness, socialist realism, the *nouveau roman,* and so forth — not to mention the intricate rules of harmony and counterpoint in music. The English physicist Dirac, a Nobel laureate, said recently: "It is more important to have beauty in one's equations than that they should fit experiment."[11] The counterpart to this is the statement by Seurat on his pointillist method: "They see poetry in what I have done. No, I apply my method, and that is all there is to it." In other words, the experience of truth, however subjective, must be present for the experience of beauty to arise, and vice versa: An elegant solution of a problem gives rise in the connoisseur to the experience of beauty. Intellectual illumination and emotional catharsis are complementary aspects of an indivisible process.

VII. Regression and Rebound

I would like to conclude this discussion with a remark which is no more than a hint, to place the phenomena of human creativity into a wider biological perspective. I have talked of temporary regression, followed by a rebound, as a characteristic of the creative act. Now biologists are familiar with a similar phenomenon on lower levels of the evolutionary scale. I mean the phenomenon of regeneration.[12] It consists in the reshaping of bodily structures — or the reorganization of functions — in response to traumatic challenges from the environment. It involves the regression of bodily tissues to a quasi-embryonic state and the release of genetic growth potentials which are normally under restraint in the adult organism — just as in the moment of discovery the creative potentials of the earlier forms of intuitive thinking are released from the censorship of the conscious adult mind. Psychotherapy reflects the same process on a higher level. It aims at inducing a temporary regression in the emotionally traumatized patient in the hope that he will regenerate into a pattern which eliminates the conflict. The creative act could be called a kind of do-it-yourself psychotherapy where the traumatic challenge is intellectual instead of emotional, for instance, new data which shake the foundation of a well-established theory, observations which contradict each other, problems which cause frustration and conflict — or the artist's perplexities in trying to communicate his experiences through the blocked matrices of conventional techniques.

And finally we find the same pattern reflected in the death-and-resurrection motif in mythology, in Toynbee's *Withdrawal and Return*, in Jung's *Night Journey*. Joseph is thrown into a well, Mohammed goes out into the desert, Jesus is resurrected from the tomb, Jonah is reborn out of the belly of the whale. The mystic's dark night of the soul reflects the same archetype. It seems to be a principle of universal validity in the evolution of individuals and cultures.

Notes

*Reprinted from *Challenges of Humanistic Psychology*, edited by James F.T. Bugental. © 1967 by McGraw-Hill, Incorporated.

1. Arthur Koestler, *The Act of Creation* (New York: Macmillan, 1964).
2. G. De Santillana, *Dialogue on the Great World Systems* (Chicago: University of Chicago Press, 1953), p. 469.
3. I owe the term "*Haha* reaction" to Dr. Brennig James's paper "The Function of Jokes" (unpublished), which he kindly sent me.
4. For a review of the theories on laughter, see Koestler, *The Act of Creation*, chaps. 1 and 2; also Koestler, *Insight and Outlook* (New York: Macmillan, 1949), part 1 and appendix 2.
5. Cf. Koestler, *The Act of Creation*, pp. 271, 284; and, for a bibliography on the psychology and physiology of weeping, pp. 725–28.
6. Ibid., pp. 285–300.
7. See T.S. Kuhn, *The Structure of Scientific Revolutions* (Chicago: University of Chicago Press, 1962); K.R. Popper, *The Logic of Scientific Discovery* (London: Hutchinson, 1959).
8. See E.H. Gombrich, *Art and Illusion* (London: Phaidon Press, 1962).
9. J. Hadamard, *The Psychology of Invention in the Mathematical Field* (Princeton: Princeton University Press, 1949).
10. E. Kris, *Psychoanalytic Explorations in Art* (New York: International Universities Press, 1952).
11. P.A.M. Dirac, "Evolution of the Physicist's Picture of Nature," *Scientific American* 208 (1963): 45–53.
12. Koestler, *The Act of Creation*, pp. 447–74.

CREATIVITY IN SCIENCE

R. HARRÉ

I. Novelty and Creativity

To create is to produce or generate what did not exist before, and most importantly, it is to produce not only an individual which did not exist before but one of a new and hitherto unknown kind. In science the most obvious product of creativity is a sort of discourse, the flow of theory. But theory is itself a secondary product, a description of potent things and products which produce the phenomena we experience. And yet, at least initially, the potent things and processes described in theory are not part of that experience. It is in our conceiving of ideas about them, by imagining possible potent things in which and among which causal activity occurs, that creativity is exercised.

But if theory is to provide understanding, it must be intelligible, and that intelligibility must derive ultimately from the intelligibility of the novel entities and forms conceived in the creative scientific imagination. So novelty must be tempered by connection with the known, or at least with that amongst the known which we take to be intelligible. What it is for something to be intelligible will emerge in the course of the discussion. But the only possible connection that would allow both intelligibility and novelty is that of analogy. New things and processes must be like known things and processes in some ways, but must be unlike them in others. The forms of unlikeness may be very various. Unlikeness may derive from the absence of some common property, as photons have no rest mass, or it may derive from a combination of a set of properties never found together in ordinary experience, as the spatio-temporal continuum must be both continuous and infinitely divisible. Sometimes, as in the latter case, the resolution of the consequent paradoxical intuitions is achieved only by proofs of consistency created in the formal domain of mathematics.

D. Dutton and M. Krausz, eds., The Concept of Creativity in Science and Art, pp. 19-46.
© *1981 Martinus Nijhoff Publishers, The Hague/Boston/London. All rights reserved.*

In this article I shall follow the creative scientific imagination in some of its acts and examine some of the constraints and disciplines which have developed to banish fantasy from the theorizing of scientists. These, we shall see, derive from stricter and stricter ideas of what is possible. And in doing this we shall be pursuing the philosophical issue of realism, since we shall soon confront the problem that the realm of the real seems to extend far beyond the realm of the experiencable.

II. False Theories of Creativity

The importance of creativity will vary with different views of science. In a simple inductivist or positivist view of science, the passage from fact to theory is achieved by a purely formal addition of generality to observed fact, and logical axiomatization of the "laws" so derived. Scientific creativity is then, at bottom, no more than formal invention. To someone taking science seriously as it appears to be, namely an intellectual enterprise in which ever-new content is added to the revelations of experience, the creativity involved in the invention of new concepts can only be an illusion of the advancement of possible experience for an inductivist. On the positivist view this burgeoning of new concepts can have at best a literary purpose, providing a kind of attractive appearance to propositional structures and linking concepts that have only logical importance as bridges from one observation statement to another. Indeed the creative act is strictly dispensable, since all that is essential to science is contained in descriptive concepts and the logical and grammatical particles that are the connectives of their structure as a discourse, if we follow the inductivist-positivist view.

According to realist views of science, the concepts of theory do refer to possible real processes in real structures of things, and thus the introduction of a novel concept enlarges the possibility of experience and adds something to our conception of the world. But realist philosophers of science differ fundamentally on the issue of whether the appearance of a new theory or of its component concepts in the scientific community, and presumably once in the mind of an individual scientist, is a process which is susceptible to rational analysis, and then to reduction of the obedient following of certain canons. Thus some realist philosophers hold a purely psychological (and thus serendipitous) theory of scientific creation. According to Koestler in *The Sleepwalkers,* scientific invention is some kind of accidental occurrence mediated by psychological processes, mysterious to the inventor himself, and amenable only to the kind of analysis he provides of Kepler's mind: part history, part biography, and part psychology.[1] But his analysis of Kepler's intellectual life lays bare certain preparations, which anyone must undertake if he is to hope for a "discovery," in the sense of the invention of a new concept bringing order to the data. Kepler *already* had the ellipse, as a form, both geometrical and analytical, before he could creatively apply it to the problem of making sense of the orbit of Mars. There is then a two-stage process, the invention or acquisition of the required concept, and then the novel use of this concept bringing out something new in the data. In just this way in microsociology, once we have acquired Goffman's concept of the "with," the symbolically displayed pairing, etc., of people in public places, the world is suddenly full of "withs."[2] And what is more, not only must the specific form of the image or concept be present to the mind of the creator, but it must inhabit a mind prepared for that kind of form, as Kepler's was prepared for *an* orbit of harmonious proportions, by his adherence to a general metaphysics, or supra-theory, according to which all the processes of nature were compounded of processes involving forms having harmonious proportions. As I shall argue in detail later, Kepler's situation corresponds precisely with the scheme for rational analysis of scientific invention which merges from the theory I shall propose. Kepler had already studied a highly specific analogue of the orbit of Mars, namely the geometrical ellipse, and was thoroughly acquainted with its properties, and with the various consequences which followed from such properties as the ratio of semi-diameters. And, from his earliest studies, he had been convinced of the verisimilitude of a world picture of the universe, in terms of which the attempt to find a regular geometrical form to correspond to and represent the orbit of Mars made perfect sense.

In the same vein, the fallibilist theory of Popper rightly lays stress on the role of invention in science and the role of the imagination in that invention.[3] But the discipline under which that faculty works, making it the imagination of a scientist, rather than the fantasy of a crank, is left unanalyzed by Popper, and assigned to the "psychological." This unfortunate move is clearly connected with Popper's very narrow vein of rationality, identified by him throughout his works with adherence to the canons of deductive formal logic. Of course the way invention occurs in science must be a topic for psychological study alone, and can conform to no schema, and have no canons of rationality, if rationality is confined to the principles of deductive logic. But one must take great care to distinguish the bogus claim to rationality of the inductivist, who purports to pass beyond experience in the dimension of generality, and the genuine claim of the realist, who sees the imagination of scientists generating conceptions of things, properties and processes that pass beyond any actual experience, not because they make some claim to universality, but rather provide an inkling of the way the world is here and now in those regions like the very small and the very distant, to which we have neither sensory nor instrumental access. The defender of the rationality of creativity seeks the canons of reasonableness in accordance with which such imaginative constructions of conceptions of the unknown can be rated proper or improper, plausible or implausible.

III. Science as Icon of Natural Structures and Powers

An adequate theory of scientific creativity can come only from a properly constituted view of what is being created. As in most issues in the philosophy of science, all will depend upon how far one regards the analysis of theories to be primarily a matter of laying bare their logical form. This article is written from very much the view that little of interest to the understanding of science and its modes of thought can be found by a search for logical form. Scientific thought cannot be understood in terms of the content-free principles of formal logic, be they deductive or inductive, if indeed there are any content-free principles of the latter sort. So if we regard theories, primarily, to be considered as formally ordered structures of propositions, we shall look only for the sources and means of creation of those structures. We have already seen how this leads to a stultifying psychologism.

In my view, a theory must be considered as it conforms to certain principles of content, and it must be analyzed for philosophical purposes, so as to bring out the various sorts of propositions it involves, classified by reference to the kind of thing they assert about the world. The logical form of such propositions and the logical structure of the discourse within which they appear is not, of course, irrelevant to our understanding of them, but, I hold, is far from exhaustively determining all that a philosopher might want to say about them.[4] Thus, for example, a causal proposition cannot be identified by its form alone, say that of a conditional, but is only truly causal if it explicitly or implicitly refers to an existing natural agent potent to bring about the causation when unconstrained and suitably activated. Thus, "ignition of petrol causes combustion" is intelligible as a causal proposition on two counts, neither of which can be dispensed with: it has the form "If i then c," and it refers to "petrol" an inflammable liquid. Our understanding of the proposition as causal depends upon our understanding of "petrol" as "inflammable," that is, as something which naturally tends to burn in natural conditions when lighted.

Let us first ask in the most general way, but with more care than is usually applied to these matters: What is the content of a theory? We shall avoid, for the moment, the few very general, very atypical theories, that one finds in fundamental physics, be it classical or modern, and stay with the kind of theory that is typical of chemistry, or medicine, or physiology, or social psychology (of the reformed, ethogenic, sort). Commonly then, in addressing a theory, we confront a discourse which seems to be describing some arrangement of things with certain definite properties, the modulations and changes of which are responsible for the phenomena we are theorizing about. It might be that the distribution and form of animals and plants, both geographically and geologically, requires explanation. Darwin and Wallace produced theoretical discourses describing a process, which, repeated billions of times, was responsible for the phenomena, as they saw them, in the light of the theory.

Already much has emerged. Notice first how readily one slips into speaking as much of an explanation as a theory in this sort of example. I shall return to this point. Notice too that the way the observations of naturalists present themselves to the great biologists, the form of order they saw in them (such as spotting the gradation of the size and form of the beaks of Galapagos finches) was a product of holding the theory. This is not just a psychological observation, and we must pause to examine it further, for here a crucial moment of creativity occurs, one liable to be overlooked in our awe in the face of the invention of the grand design.

Recall, if you will, the discussion of the inductivist view of science and the narrow margin of creativity it allowed, doing science simply being the generalization of regularly recurring observational correlations. Mill, the hero of the inductivist view, describes this process as generalizing over similar cases. The important point to notice here is that he takes similarity for granted. The development, skeleton by skeleton, from eohippus to horse, is similar, for evolutionists, to the development, beak by beak, of one species of finch into another. But why should they be regarded as similar? And in what respects? Their similarity for evolutionists is problematic, and needs to be explained. Taking it for granted, Mill sees scientific method as imposing only a purely formal concept, generality, upon our scientific knowledge.

We must turn to the hero of neo-Kantian realism, Whewell, for a resolution of what is problematic in the inductivist view.[5] For Whewell, scientific method is creative not only as to logical form, generality, but also as to content. A science, he holds, is produced by the existence of an "idea," an organizing conception, which is brought to the phenomena, and creates both the possibility of perceiving similarities, which bring the phenomena into the same class, and supplies the generalities, all at once. In controversy with Mill, he cited Kepler's discovery of the orbit of Mars, involving the ellipse, not as a generalization of the known positions of the planet in the star charts, but as a prior idea which provided organization to those positions (in some cases this may be sufficiently sensory to be called a gestalt) under which they became moments in the career of a planet along an elliptical orbit. So the gradation of the beaks of finches is both a product of, and evidence for, the idea of continuous evolution of species one from another.

But the Darwin-Wallace theory was not just the idea of gradual but accumulated change, the idea of organic evolution, which had been held in other forms prior to their development of it. Their theory purported to describe a process occurring over and over again in nature. But neither had observed this process; indeed, it has scarcely been observed today, outside the glass walls of drosophila boxes. (Ford's butterflies and industrial melanism are recent exceptions.) How could Darwin and Wallace describe the process of evolution in such detail if they had never, and given the time span of the process never could, observe it? What were they describing? What stood between them and the almost untold aeons of minute changes in millions of species on millions of hectares of the earth's surface that they reduced to order? Clearly they shared an imaginative conception of the organic history of the earth and the natural forces and processes that shaped it. I am not interested, for the moment, in the question of from whence they derived that conception, but in the role of the conception itself.

It is this conception which stands between their limited experience of organic biology, and the utterly out-of-reach organic history of the earth. This again is not just a psychological observation about the personal thought forms of a couple of Victorian giants. By looking very carefully at the form such conceptions commonly take, we shall see how the necessary intermediary between ignorance and the unexperienceable is created, and so how theories can come to have content beyond the description of experienced phenomena. The conception which lies between ignorance and reality I shall call an *icon* of that reality. It is not a delineation since that reality is usually not known, though it may come to be known. In general, a theory describes an icon as a representative and surrogate for that reality. I choose the word "icon" in preference to the more commonly used "model," since the latter has long ceased to be univocal. I also want to draw attention to the frequent sensory or imaginative character of the bearers of conceptions of unexperienced reality, though as we shall see these develop away from the sensory into abstract forms. Darwin and Wallace "formed a picture" of the organic process, a picture which by their description of an icon of the organic process they convey to us. Of course, psychological idiosyncrasy is such that no psychological generalizations about how their conception of nature was present to their minds or ours, their readers, is intended. In particular, I do not intend to suggest that they or we must literally visualize the icon of reality at the heart of a theory. But by keeping a sensory connection, by the use of this word, I want to emphasize that nature must be conceived as a process in time, and a structure in space, of individual organisms, geographical and meteorological forms, ecological interrelation, and the like. Darwin and Wallace attempted to conceive of the reality and to convey their conception to us, though neither they nor we can experience that reality.

What then of the supporting evidence that these great men cite in their works? What relation does it bear to the conception of unexperienced processes responsible for all the phenomena of organic change? We know from the arguments of the Humean tradition that it can provide no logical support for the generalization of the theory or for the claim of the theory to universality within its domain. To understand the role of evidence in science we need to take a radically different view of it, more radical than Popper's fallibilism. I shall try to show by examples that a great scientist cites supporting evidence, not as premises or even as evidence in the legal sense, but as anecdotes, illustrative of the power of the theory to make certain widely-selected phenomena intelligible. Conceiving of the citation of evidence as anecdote brings the explanatory power of the theory to the fore, and raises the philosophical problem of what it is to make the phenomenon intelligible. So Darwin's account of the gradations of beak shape among finches from different islands in the Galapagos group appears not as a premise from which his theory might be inferred, together with all the other available evidence, but rather as an anecdote illustrative of the power of the icon to make the phenomena intelligible.

This can be shown in other cases in just as striking a fashion. In a currently influential work, *Asylums,* Erving Goffman sets out a theory of institutions, based upon the idea that an institution should be conceived both as a device for fulfilling its official functions, i.e., as a hospital is a place where the staff cure people, and as a setting for the staging of dramas of character, where personas are created and defended, so that a hospital is a place where people perfect dramatic performances as "surgeon," "nurse," and of course "patient," learning to conform to and excel in these dramaturgically conceived "roles."[6] In his discussion of those closed institutions he calls "asylums," Goffman cites instance after instance of people doing things that become intelligible only if conceived on his dramaturgical model, his icon of the nunnery or the barracks or the hospital as a theatre. Each citation is an anecdote, in which the power of the dramaturgical theory to make phenomena intelligible is illustrated. At the same time, it becomes plain that certain structures and textures of life in such places become visible, stand out from other phenomena, only if that life is examined by someone with that icon of the institution in mind.

I can sum up this theory of science in the phrase "icon and anecdote," bound into a single discourse by the explanatory power of the icon, its power to make our experience of the world intelligible, since above all our scientific icons are depictions of the productive processes which bring the patterns of phenomena into being. We must also keep in mind that our experience is only wholly what it is when we conceive of the world that way. Thus, understanding its genesis makes the ordering of experience in classificatory systems intelligible. In science, for every phenotype we find convenient to extract from experience, we must conceive a genotype, for every nominal essence we use in practice to select and identify things and samples of materials, we must conceive a real essence. I was pleased to read in Lévi-Strauss the following elegant statement of the icon and anecdote theory.

> . . . Social science is no more founded on the basis of events than physics is founded on sense data: the object is to construct a model and study its property and its different reactions in laboratory conditions in order later to apply the observations to the interpretation of empirical happenings which may be far removed from what has been forecast.[7]

Creativity, then, must be at its most seminal in the origin of conceptions of the unexperienced, icons of the reality beyond but productive of our experience.

But why not just study the discourse, and never mind about the icon? First, it is *assumed* that a discourse can contain all that is in the object of the discourse, so that all relations relevant to the internal structure of the discourse are supposed to be somehow present within it. But there may be relations in the discourse, relations which we know and use, say between predicates, where two predicates permeate each other's sense, or relations of analogy between predicates deployed in an argument, which are dependent on prior icon relations, appearing in the icon as coexisting properties, or likeness and unlikeness between things. Thus, the copresence of a pair of properties in an icon becomes the source, in a diachronic process of meaning assimilation, of synchronic internal relations between two predicates, relations which may be quite crucial to understanding how the discourse is structured and how its development would be justified. Such relations can never be reached by the analysis of a discourse into its logical form, since a fortiori logical form extracts from all material relations between predicates, and yet it may be just those relations upon which the coherence of the discourse as a scientific theory rests.

IV. Two Epistemological Regions

"Beyond our experience," you say. "But surely traditional epistemology teaches us that we have no knowledge of what is beyond our experience." There are two epistemological barriers which scientists regularly overleap in their practice. The one is that erected by positivists between actual and possible experience, the other by the critical philosophy between all actual and possible experience and the realm beyond all possible experience. I shall try to show that our powers and techniques of creativity are such that we can, at first tentatively and cautiously, and finally boldly, trust ourselves to pass beyond them, and dwell in thought, in worlds accessible only to our creative imagination.

The realm of actual experience is limited in two ways. We are confined to the senses we actually possess, and invariants we can express between them, so hand, eye, and ear conspire to provide our experience of a bell. But without conception, perception is blind. And what we can actually identify within and across our sensory fields depends upon the sort of concepts with which we are mentally prepared for the world. As Kant had it and contemporary psychology confirms, our experience is a product of schematic ordering and supplementing our sensations, the schemata being of conceptual and perhaps even of a linguistic origin.[8] All this is commonplace. But it is not difficult to imagine things, and structures and processes and properties, that are too far, too small, too fast, or too slow, or even too big to be experienced by us as we are presently constituted, though the Gibsonian invariants in the objecthood of these entities are just the same invariants as in the objects of ordinary experience. Locke talks of what we might see with microscopic eyes, or hear with a more acute sense of hearing, and Geach has speculated on the colors we would see were our eyes sensitive to a broader spectrum of electromagnetic radiation. In this way we can conceive of a realm of possible experiences, and populate it in our imagination with objects undergoing processes which we do not, but might, experience.

But theories describe icons of structures and processes which would be experienced within that very realm. Kepler conceived of a fine structure for the snowflake which would be a natural form of packing for minute ice particles (water molecules, if neat little spheres) and which would, if much repeated, yield the universal hexagon. Kekulé conceived of a structure in space in which the carbon atoms of benzene would form a stable ring, and Harvey completed his sanguinary plumbing with minute imagined vessels, closing the hydraulic circuit, while Van Helmont conceived of disease as the invasion of the body by an army of invisible minute organisms. These men populated a world of possible experience with fabulous creatures of their imagination just as Mohammedans filled the air with djinni, Descartes ignited a furnace in the heart, and Velikovsky supposed a radically different history for the earth. So entering a realm of possible experience in the imagination is fraught with hazard, for the imagination is as capable of fantasy as of sober speculation. I shall return to examine the discipline to which it is subjected in science and from which we shall extract the fragments of a creative criterion. It is because of the existence and acknowledgment, and ultimate justifiability, of this discipline that we can override the extremes of positivism which would have us conceive of science as no more than the "mnemonic reproduction of facts in thought." Let us call the activity of thought in populating the realm of possible experience the work of the reproductive imagination.

But the scientific imagination does not confine itself to the same realm of creation as is continuous with the realm of perception. It attempts to conceive of the structure of the world beyond all possible experience. Scientific thinkers are driven to attempt this ultimate barrier to knowledge by two factors. The first is simple. Our ordinary experience is full of instances of phenomena whose effects are inexplicable by any work of the reproductive imagination. Electric and magnetic phenomena are the most striking examples. We are simply not equipped with sense organs sensitive to the magnetic influence.

First attempts to solve the problem of the mechanism of the magnetic influence involved the work of the imagination at its reproductive stage, leading to a proliferation of magnetic fluids, particles and the like, clearly denizens of the realm of possible experience. Norman and Gilbert went further. Gilbert's imagination leapt the barrier of all possible experience, leading him to postulate the *orbis virtutis,* a shaped, structured field of potentials or directive powers. A structured field of *powers* was something, though spatially extended and temporally enduring, that was clearly not, as such, an object of possible experience, given our lack of magnetic sensibility. Eventually Faraday offered to the visual sense a picturable icon in the lines of force technique for representing the field.

Not only is the perception of the magnetic influence beyond the range of any of our sensory fields, but magnetism, like electricity and gravity, forces the creative imagination to the transcendental stage, since there is a conceptual, not just a contingent, difficulty about the perception of fields. A field is a distribution of potentials, and though we speak of the energy of the field at a point, that energy is not manifested in any kind of action. Thus a field is *a fortiori* imperceptible, its existence known only from its manifestations, and from the presence of field generators, like conductors carrying a current, iron atoms, lumps of matter, and the like, the laws of structure of whose produced fields we have discovered by examining the manifestations of the field on other similar occasions. The icons which represent fields, like rotating tubes of moving, elastic fluids, are not representative of objects or processes in the realm of possible experience, and to suppose them to be so would be a major epistemological error.

But the matter is more complicated. The transcendental imagination is required to generate not only distributed potentials, but to conceive an intermediary between potential and action, not as far beyond possible experience as the potential, but in its realm. At some definite point in the room there is a gravitational potential. But at this moment no test body is at that point. Suppose we now bring a test body to that point, say an alabaster egg, and support it there on a platform. There is still no action, but since action is now possible if the platform is removed, we are obliged to postulate another entity — the tendency to fall acquired by the alabaster egg from the gravitational field. We know that the egg acquires the tendency from the field, since on the moon, for example, we have good reason to think that the egg's tendency to fall will be much diminished, so the tendency is not an intrinsic property of the egg. And we can easily check whether the egg has indeed acquired the tendency by removing the platform and seeing if, indeed, the egg does fall. If we replace the platform by a hand under the egg we might claim to be experiencing the tendency of the egg to fall, but we could hardly claim that it is an experience of the potential of the field, since that potential will be there when the egg is removed, and then we feel no tendency for downward fall. The tendency, then, is not so far from the edge of the boundary of all possible experience as is the potential, and may indeed be held to within the realm of possible experience if the experience of weight and pressure to which we apply our powers of resistance, frustrating action, is accepted as the experience of a tendency, and I cannot think of any argument that would oblige us to deny this.

Contemplating the egg has led us to a complex icon of the whole situation, and in supposing it to be a delineation of the real, the subsequent fall of the egg becomes intelligible, as does the dent it makes in the velvet cushion upon which it usually rests. There are two material things, the earth and the egg, and two immaterial things, the gravitational field, our conception of which is certainly a product of the imagination in its transcendental phase, and the tendency caused in the egg by the power of the field, a thing arguably on the border of the realm of possible experience.

V. Two Acts of Creative Imagination

I shall try to show that though the creative imagination of scientists is, in a certain sense, free, and indeed we shall come to see in exactly what sense, nevertheless, analytical schemata can be constructed to represent the dynamics of concept-construction, in an idealized form, and from which canons, exemplified in rules, can be abstracted. The essential point to be grasped, in considering the acts of the imagination in its reproductive phase, is that in producing an icon of a possible reality, the imagination is not modelling something known, but something which is in its inner nature unknown. We know how the mysterious structures and things behave, since they have produced the patterns of phenomena we wish to make intelligible, so at least our icon must depict a possible being which behaves analogously to the unknown real being.

But simply to conceive possible realities in terms of their behavior is no advance on positivism, since we already know the behavior patterns of things and express these in the laws of nature. The task of the reproductive imagination is deeper, for it must enable us to generate a conception of the nature of the objects which behave exactly like, or in various degrees analogously to, the real things actually in the world. The forms of thought involved in this act of the reproductive imagination can be idealized and schematized by distinguishing between the subject of the conception, what the icon is a model *of*, and the source of the conception, what the icon is modelled *on*. Since the nature of the subject is unknown in the sense of beyond all actual experience, the relation between subject and icon can be mediated only by likeness of behavior: a swarm of molecules behaves like a real gas behaves, a person in a shop behaves like an actor playing "person in a shop," the evolutionary process produces results like a plant breeder produces. But the creative act of the reproductive imagination is to produce an icon of the unknown nature of the real world, and this icon must be at least capable of being recognized and understood as a plausible depiction of a possible generating mechanism for the patterns of behavior whose explanation is our problem. So if we imagine an evolutionary process as consisting of minute variations in form and function generation by generation, and certain of these variations leading to greater reproductive rates in their possessors, we have conceived a mechanism (or at least the rough outlines of a mechanism) which would generate the pattern we call "the evolution of species," which is itself a product of an act of structuring upon the individual bits of knowledge about form and structures of individual organic specimens.

But the natures of things imagined and the generative mechanisms they severally constitute must be plausible, both as generators of the observed patterns and as possible real existents. Thus, not any sources will do. Sources of models, as icons of reality, must conform to two criteria:

1. They must be the kinds of things, structures, processes, and properties the current world picture regards as admissible existents — gases rather than imponderable fluids (1800), material atoms rather than atmospheres of heat (1850), electric charges rather than solid atoms (1920), neural networks rather than mental substance (1960), and so on. The metaphysics of science consists in the discussion of the coherence and plausibility of the world pictures, literally conceptions of structures, that occupy space and endure for a time, and any possible systems they may form, which serve as general sources for conceptions of possible realities, though, in one of the examples I have cited, that of elementary electric charges, we are on the borders of possible experience.

2. Having thus conceived proper kinds of things, we need to be able to imagine plausible laws for their behavior. And we find such laws, or closely analogous laws, in existing science. Electrons obey, *for us,* Coulomb's law, the law obeyed by the charges on suspended pith balls. Gas molecules obey Newton's laws of motion, the blood in Harvey's imagined tubules obeys the ordinary laws of circulating fluids. It is thus that the disciplined imagination works in reproducing a version of experienced reality in the realm of possible experience. This kind of representation may, in certain cases, turn out to be not just a representation to fill a gap but a depiction of reality itself.

But what about the human imagination in its transcendental employment, its transcendental phase so to speak? The first point to notice is that the behavioral constraint is the same in both phases. The world as conceived beyond all possible experience must behave just as the real world behaves, or in a very similar manner. So the dispositions we assign to reality in our imagination must be closely analogous to the dispositions we find the real world to have. But if these dispositions are to be grounded, that is, to be powers and liabilities, dispositions grounded in the nature of things, must we not try to conceive of natures of things the details of which *must* lie beyond the boundary of all possible experience, and if that nature is beyond all possible experience, how are we to conceive it? Yet it is my contention that physicists, cosmologists, and psychologists both can and do achieve creative acts of the imagination in the transcendental phase of activity, and that we can follow them. How is this possible?

The first point to notice is that the world beyond all possible experience can share one kind of attribute with the objects of the world of possible experience, namely structure. It is this attribute that makes for the possibility of intelligibility of conceptions of that world, and of course for its mathematical description. Objects in the experiencable world must have both sensory qualities, or considered in themselves, at least the power to manifest themselves sensorily, and they must have structure (structure which need not be, though it often is, spatial), while processes in that world, to rise above the dead level of imperceptible uniformity, must have both powers of sensory manifestation within thresholds that allow us to say they have structure in time. (Compare the difficulty that was once experienced in knowing whether the Australian aborigines had melody, when the multi-toned structure of their tunes is within a semi-tone, our normal unit of melodic differentiation.) In the world beyond all possible experience, there are, *a fortiori,* no powers of sensory manifestation, though two connections must remain with the world of experience. In the one there must be an actual causal connection, in that the objects in that world must have powers to affect objects in such a way as to change or stimulate or release their powers to manifest themselves or changes in themselves to us. And secondly, more germane to the issue of conceivability, both synchronic and diachronic structure may be attributable. For example, in Medieval chemistry, the four principles which had the power to manifest themselves in warmth, coldness, wetness and dryness, though not perceptible themselves as such, nevertheless were imagined to be present in things and materials in definite proportions (non-spatial synchronic structure) which might be changed, and so change the nature of the substance, a process, if intelligible, having diachronic structure.

The role of icons in conceiving qualityless structure thus becomes both clear and shows itself to be equivocal: for icons, if based upon the reproductive power of the imagination, but passing into the transcendental phase, will constrain conceived structure to the structures of possible experience. The imagination, in its transcendental phase, must proceed to acts of abstraction and generalization to pass beyond this constraint, and of course the acts of the imagination in this phase are identical with abstract mathematical creation. In the very last analysis icons of the world beyond all possible experience *may* be required to have the character of abstract mathematical structures. But of what?

We have already noticed how, in science, dispositions are grounded in hypotheses about the natures of the individuals which manifest these dispositions. If the individual is a source of action such as a material thing as the ultimate source of the gravitational field, or an acid, relative to chemical analysis at the molecular level, then the grounded disposition is a power. But if, as we move into the transcendental phase, we have left behind all properties other than structural with which we might ground dispositions, then those structures in which all secondary dispositions are grounded must be of primary or pure dispositions, that is, of ultimate powers, that is, in the elements of the most fundamental conceivable structures, powers, and dispositions must coincide. In natural science, fields are the best example of structures of pure powers. Icons, such as Maxwell's tubes of fluid, may be required for conceiving structures of potentials, but here at least they are but dispensable models of our abstract conceptions of reality. They are, at best, a system of metaphors for holding onto the sense of the abstract objects, and it is to this role that another aspect of the sense of the notion "icon" is directed. An icon as a religious painting is not just a picture of some worthy person, but is a bearer of meaning, generally abstract with respect to that which it depicts. In their transcendental employment the models generated by this phase of the imagination are truly icons.

VI. Disciplining the Fantasy

Common sense would have it, no doubt, that the test of the imagination in conceiving objects in the realm of possible experience is whether, when our senses are extended by the development of some device, such as the stethoscope or the microscope, the hypothetical object or process appears, that is, it is a matter of whether and to what extent reality, when it is revealed in experience, matches the icon. But common sense needs defense. There is a tradition in philosophy of casting doubt upon the authenticity of what is perceived, by insisting that only the existence of and properties of the immediate elements of various sensory fields involved are known for certain. Happily, as far as stethoscopes, probes, microscopes, telescopes, and slow motion film are concerned, one can establish a gradual transition from the objects and processes of unaided perception, to the sounds, shapes, colors, motions, and so on, brought into our experience with the help of instruments. One can hear the same sound with or without the stethoscope, but one can also hear clearly sounds heard only faintly or not at all without it. Thus, we establish a continuity of the existence of percepts. By this achievement, the world of possible experience penetrated by instrumental aids is made one with the world of actual experience, so the extended world is no more nor less dubious or inauthentic than the world of unaided perception, the ordinary world. And this is all we need for the control of creativity at the reproductive phase of the work of the imagination. Philosophers may continue to argue about the epistemological and metaphysical status of material objects, but their disputes and distinctions cannot detach bacteria from bodies, nor galaxies from ganglia.

But discipline in the world imagined to lie beyond all possible experience cannot be based wholly upon instruments. However, there is a kind of penumbral region, wherein structure is simply spatial, where structures whose elements are beyond all possible experience may nevertheless be displayed. I have in mind the photographs of molecular structure obtained by field ion microscopes, or the tracks of "particles" observed in cloud chambers. While the phenomenal properties of the structure are linked to its elements only by long and sometimes ill-understood causal chains, the structure so projected is at worst isomorphic with the structure of the thing or process being examined, at best that very structure itself. However, if we consider cases deeper into the inexperiencable, only reason can come to the aid of the creative imagination, and that only *a posteriori*. At the deepest level, the best that we can do is show by argument that the structure of elementary powers we have imagined as the ultimate structure of the world fulfills certain necessary conditions for the possibility of our having the kind of experience we do have, and there may be a still more general form of argument which would link certain structures (and certain powers) to the possibility of any experience at all. In fulfilling these conditions, the world as we experience it is made intelligible.

A process or structured object becomes intelligible if the following conditions are met:

1. From the imagined fundamental world structure the form of the process or object can be deduced, i.e., from the tetrahedral distribution of the valencies of the carbon atoms the observed form of the diamond can be deduced with the help of certain ancillary hypotheses, that is, the structure of the valencies provides a reason, *relative to accepted physics,* why diamond has the form it has. Form, as we may say, is inherited from form. The intelligibility of the form of diamond comes not just from the fact that the proposition expressing this stands in a certain logical relation with some other propositions, but that among those other propositions are some descriptives of some underlying and fundamental form, that is a structure of units or elements that are, for that case, not further decomposable. Thus, to cite structure is to make intelligible, and by linking structure via the deductive link, which has the effect of preserving content, that intelligibility which derives naturally from citation of structure alone, is transferred to the form of the diamond. But sensory qualities, like color or timbre, cannot be so made intelligible, only their associated forms, wavelength, or harmonic structure, can be referred to more fundamental forms and so acquire intelligibility. It is a prime rule of science that qualitative difference should be explained in terms of structural difference, and in so doing, the only final form that an explanation could take is achieved. In short, only form or structure is intelligible in itself. Philosophical argument

38

for this proposition could do no more than take the form of the analysis of all satisfactory explanatory forms and the exhibition that their satisfaction derives from that feature.

2. In the nodes of the imagined world structure, there are agents, that is the structure is a structure of sources of activity, as for example a complex formed from repressed traces of disagreeable experiences, a concept which represents a state of the world beyond all possible experience, is a structure of agents, its elements having power to effect changes in behavior, so that the complex itself becomes a structured agent, the source of the pattern of neurotic or compulsive behavior, making it intelligible by showing that the form of the behavior is a direct transformation and manifestation of the form of the complex.

VII. Society as Created Icon

But when we create icons in the pursuit of the social sciences, we cannot take it for granted that there is a real structure, some independent world, of which that icon, however imperfectly, is a representation. Indeed, both the existential status of society and the significance of societal concepts is highly problematic and cannot be taken for granted. We speak of the nation, the army, the middle class, as we were speaking of the island, the Thames valley, and so on. Our power to create societal concepts is, as we shall see, a creativity of another kind.

I shall approach this difficult problem through two examples, illustrating different facets of the role of societal concepts, and their associated icons in our lives. Both examples will illustrate how we are unprepared to live in an unintelligible environment, that is, an environment which does not either exhibit structure or clearly manifest an underlying structure. Imagine a large complex of buildings unified by a boundary wall, and a common calligraphy in the labels displayed at various entrances (that a hole in a wall is an "entrance" is also a social, not a physical, fact, so in this analysis an underlying and unexamined ethnomethodology is being taken for granted, but at least by me, knowingly). People move in and out, some prone on stretchers, others arrive with every mark of respect in Rolls Royces. Inside, uniforms are much in evidence. Many people are in bed, and even some of those who are walking around are wearing pyjamas and dressing gowns. What on earth is this strange place and what is going on? The innumerable momentary interactions and sayings of individuals, or rather some part of them, are made immediately intelligible by the hypothesis that this is a *hospital*, that is, the whole begins to exhibit structure. The introduction of this concept is strictly comparable to, though much more complex than, the introduction of the concept "galaxy," which made the appearance of the night sky intelligible by referring its observed form to an underlying and aesthetically pleasing structure, the spiral form of the stars in the galactic plane. As ethnomethodologists have insisted, there is *always* an everyday problem of intelligibility, which it behooves sociologists to contemplate, since it is nearly always solved by those involved. Sometimes the continuous everyday solution needs supplementation by a stroke of scientific genius, as when Goffman made us realize that many flurries of activity, unintelligible within the official theory which glosses "hospital" as "cure-house," become intelligible within a single supplementary theory in which the institution is glossed as a setting for dramas of character. Each theory generates a rhetoric, a unified theory of explanatory concepts, with an associated grammar (in which rhetoric do we put the socio-grammatical rule that the superintendent can refer to the hospital as "my hospital," and to whom?). Rhetorics are drawn on in accounting sessions, in which in the course of talk the momentarily mysterious is made intelligible by allowing itself to be so described as to find a place in a structure, this time of meaningful activities within semantic fields recognized as legitimate in the rhetoric. Finally, one should notice that the official theory may find expression in what is literally an icon or icons, diagrams on suitable walls, in which structure, as officially conceived, is laid out. Nowhere on such charts appears such power and influence structures as that of the janitor in a school, where the boiler-room society over which he provides is the apex of the counter-hierarchy, and in which such officially-defined figures as the headmaster carry very little weight.

The position implicit in the example above, which it should be clear is not at all the same as the old theory of methodological individualism, can claim at least one systematic exposition in the past, namely, that of Tolstoy in the sociological chapters in *War and Peace*. As a mark of my admiration for his formulation of the theory, I have called it the "Borodino Theory," since he broaches it explicitly in his analysis of that battle and in his recurring theme of the contrast between the manner of generalship of Napoleon and of Katusov. As Tolstoy sees it, the Battle of Borodino is a middle-scale social event within an inexplicable, very diffuse, and very-large scale human movement, the periodic movement of very large numbers of people from West to East and East to West. This migratory oscillation has no name, and is not mentioned in historians' accounts of the affair. They are concerned with nations, armies, generals, governments, and the like. And their role, according to Tolstoy, is to impose order and intelligibility upon meaningless eddies in the groundswell, such eddies as the Battle of Borodino. The battle becomes, in *War and Peace,* both an instance of Tolstoy's theory, an anecdote showing its power to make phenomena intelligible, and a kind of model for the analysis for all middle-scale human events.

The battle is joined by accident in conditions which prevent either commander getting a clear view of the battlefield and its changing dispositions. Both commanders are surrounded by eager staff officers and constantly receive messages from the officers on the field. But by the time the message, usually in garbled form, has reached the commander, the situation it described has changed. Napoleon nevertheless issues detailed orders throughout the day based on the "information" he receives – but these orders rarely reach their destination, and even more rarely are intelligible to the commander they are intended for, and even more rarely still enjoin courses of action still possible by the time they arrive. But on the French side, a great flurry of command goes on. Katusov, on the other hand, does nothing, he believes no messages, he issues no orders. He waits for the issue to be decided. But as a final irony, no issue is resolved, at least in the final dispositions of that battle, though as Tolstoy points out the French losses that day turned out to be fatally weakening to their army.

But, asks Tolstoy, what do historians make of that battle? They *impose* order upon it. They represent the haphazard movements on the battlefield, enjoined by the exigencies and impulses of the moment as splendid tactical moves, flowing from the genius of the commanders, and brought about by their orders. "The Battle" as a structured, ordered, hierarchical social entity, is a product of retrospective commentary; in the technical language of the new sociology, it is an account. A series of flurries, intelligible as the mutual actions of individual men at the microlevel, become elements in a larger structure, by an act of the creative imagination. In terms of this, commentaries upon and explanations of actions are contrived for happenings which are no longer conceived as closed entities, but as elements in a larger structure having relations with other elements of that structure, for example, with the thoughts and orders of the commanders. But that larger structure has its being only in the imagination of those who share the theory, a theory, of course, which any member can hardly fail to share. Only one who follows at least one step on the phenomenological path, one who, like the ethnomethodologists, wishes to subject the natural attitude and its products, "battles," to scrutiny, can come to this. Ironically, it is in the social sciences that the positivist theory of theoretical concepts has its only plausible application, since in the social sciences the "Borodino Theory" would counsel us to treat societal concepts as serving only the interests of an imposed intelligibility, and not being referential terms pointing outside the theory to real existents.

The creative imagination of the social scientist is the most potent of all, for he can create an icon whose close simulacrum of a real world is so potent that people will live their lives within its framework, hardly ever suspecting that the framework is no more than a theory for making the messy, unordered flurry of day-to-day life intelligible, and so meaningful and bearable.

VIII. Evolutionary Epistemology

But sciences and societies have a history. And the question as to why a particular form *appears,* makes itself visible in various manifestations, at a particular time and in particular circumstances, must be tackled. To get clear on the basis for a diachronic analysis, one must distinguish the productive process of a "next stage" from the sequence of those stages. Only by clearly separating them can the problem of their several intelligibilities be solved. In general, I would wish to claim that just as in the sequential stages of plant and animal life there is no pattern from which a law of those stages can be inferred, that is, they have no intelligibility as a progression, so there is no pattern in the sequence of stages of sciences or societies. Patterns *are* discerned and described, but I would wish to argue that these reflect current ethnographies and current obsessions — God's will working itself out in history, economic determinism, and the like — the projection of which on the sequence of stages is the source of historicism. But that does not mean to say that the process of historical change cannot be understood, and that it cannot be made intelligible. I follow Toulmin in his claim that the *general form of all historical explanations* was invented by Darwin and Wallace, a form which allows for the intelligibility of a historical sequence without falling prey to historicism.

The form of our understanding of the diachrony of social and scientific creativity will be evolutionary in the natural selection mode. Thus the origin of new forms, be it animal, vegetable, or structures in thought, will be taken to be (relatively) random, with respect to the environment in which those forms will be tested. Thus, in the moment of inception, all novel forms will have the character of random mutations, and thought forms, fantasies, will be taken to be innumerable. We shall return in a moment to the important issue of how far the inception of thought forms is disconnected from their environment, and we shall find that it is not quite so clearly free as organic mutation.

But by what sort of environment are they selected? We must acknowledge the complexity of that environment. New ideas are contemplated, deliberated upon by people, and in the course of these deliberations are accepted or rejected, or sometimes merely forgotten, or abandoned because of the appearance of a novelty more in fashion. Sometimes they are tested, as to what further intelligibility they lend to what we think goes on, and sometimes even as to what they lead us to think there is. Sometimes ideas are rejected out of hand, as silly, threatening, "unintelligible," obscene, and so on. How is some order to be brought to this multiplicity?

The credit for the introduction of the basic idea of evolutionary epistemology must go to Popper, and hindsight, I feel sure, will regard this as his great contribution to philosophy. Effectively, Popper proposed to bring order to the selecting environment by the use of a principle of formal logic.[9] The instrument of natural selection upon ideas whose appearance is serendipitous with respect to that environment, and of only psychological interest, is the principle of *modus tollens,* that a proposition which has false consequences is false. By itself, "falsification" is just a logical principle, but in Popper's works it is uncritically coupled with an epistemological principle — "rejection" — that is, whatever is falsified must be rejected as knowledge. Popper's particular version of the evolutionary natural selection theory comes to grief on that coupling, since it cannot be taken for granted and it turns out to rest on two levels of theory, one metaphysical and the other scientific.

In order to pass from falsification to rejection, one must suppose that the falsified principle, hypothesis, or theory is not worthy to be accepted as knowledge. This requires recourse to an assumption about the stability of the universe, a metaphysical assumption that the universe will not so change in the future as to behave in such a way that the falsified principle is then true. But this is a principle to which Popper cannot have recourse, since it is a form of the general inductive principle of the uniformity of nature, the negation of which leads to an evolutionary epistemology for science in the first place.

But even if the passage from falsification to rejection be granted, say, as a principle, itself a mutation surviving in a hostile environment by virtue of its power to make scientific method intelligible, the application of the principle depends upon assuming some embracing scientific theory as true. In general, falsification is itself an interpretation of a yet more fundamental relation, namely contradiction. The product of a "testing" is a contradiction. "All A is B" is in contradiction with "This A is not B." To assign "false" to the principle, to make this *its* test, requires that we assign "true" to "This A is not B," that is, the assignment of a truth-value to the general proposition depends upon a prior assignment of a truth-value to the particular. That is what makes this a *test* of the general proposition. But it is notorious that the result of any experiment is very far from being a "brute fact," and we might as easily have assigned the truth-value the other way. In practice, "This A is not B" gets "true" or "false" in priority because it is embedded in a more embracing or otherwise more attractive theory than "All A is B," which is, for the purpose of the test, detached or isolated in some way. Thus the passage from contradiction to falsification is not unequivocal. (Popper did, of course, attempt a "basic statements" theory to anchor truth somewhere, but has been forced to relativize it, which is to give it up.) It seems that Popper's own attempt to give body to the general theory of which he was the originator is too much dependent on logicist assumptions as to what is rational, and on the uncritical acceptance of the transitions contradiction to falsification, and falsification to rejection.

Toulmin, by contrast, is prepared to include a much wider range of items in the selection mechanism, and furthermore, makes an important and interesting concession to a mild teleology, unthinkable in the Popperian theory.[10] The appearance of a mutant idea, for Toulmin, is not wholly detached from what has gone before in the realm of ideas, not unconnected with the tests and trials to come. As we get the idea of the kind of tests an innovation is to face, we censor what one might call "first fantasy," so that only plausible ideas are offered for the community to test. And synchronic creativity, as I have described its structure in Part I, involves a disciplinary feature in the very process of creation itself, which "close-couples" the new creation to the old. Mutation occurs, then, within a very narrow range.

What of the selection mechanism itself? Institutional and social factors become prominent once we move away from a simple logicism. Clearly, an idea will have a better chance of discussion and consideration if it is proposed by someone in a certain place in the institution, be it the society of scientists or the board of a company or the general meeting of a commune. And greater effort will be made to make the world conform to the idea before it is rejected. This latter feature is very prominent in the natural history of political ideas, where the "world" in which the idea will run its course is a human construction which can, within certain limits, still unknown, be reconstructed so as to preserve the idea unrefuted.

Toulmin's particularization of the theory is not without difficulties as well. He chooses to discuss the problem, not in the
propositional ontology of Popper, but with concepts as his individuals. However, this leads to considerable difficulties and un-
clarities in his statement of the theory, occasioned by the problem
of individuation of concepts. The problem is central, since he
treats "population of concepts" as strictly analogous to "population of organisms." But what is the "individual" concept? Is it
the concept of an individual person, so that my concept of the
atom is a different concept from yours, even though our concepts
may match throughout their semantic fields? This would seem to
be the natural interpretation, so that a concept would reproduce
itself by being more and more replicated in the minds of others.
But he also speaks sometimes as if it was the concept we shared,
which is the individual which is naturally selected, and its progeny,
not as its replicas in other people's minds, but the further logical
and conceptual descendents that it spawns. It seems clear that this
second interpretation must surely be a mistake, and that one must
stick to the individuation of a concept as my concept, and apply
the species notion to link, under the same phenotype, my concept
and yours when they are, as concepts go, alike.

I hope I have said enough to indicate that both the diachrony
of theories and societies can be understood in a general way
by the strict application of the evolutionary analogy and the
idea of selection, but that the final account of the balance between
rational, societal, and other factors in the selective environment
has yet to be struck.

Linacre College
Oxford University

Notes

1. Arthur Koestler, *The Sleepwalkers* (London: Butterfield, 1963).
2. E. Goffman, *Relations in Public* (New York: Harper and Row, 1972).
3. K.R. Popper, *Conjectures and Refutations* (London: Routledge and Kegan Paul, 1963).
4. Rom Harré, "Surrogates for Necessity," *Mind* 82 (1973): 358–80.
5. See W. Whewell, *The Philosophy of the Inductive Sciences*, new edition (London: Cass, 1967).
6. E. Goffman, *Asylums* (London: Penguin, 1961).

46

7. C. Lévi-Strauss, *Triste Tropiques*, trans. J. and D. Weightman (London: Cape, 1974).
8. See J.J. Gibson, *The Senses Considered as Perceptual Systems* (London: George Allen and Unwin, 1968).
9. K.R. Popper, *Objective Knowledge, an Evolutionary Approach* (Oxford: The Clarendon Press, 1972).
10. S.E. Toulmin, *Human Understanding*, vol. 1 (Oxford: The Clarendon Press, 1972).

EVERY HORSE HAS A MOUTH:
A PERSONAL POETICS*

F.E. SPARSHOTT

This essay does not seek to anatomize the creative process, but looks at the credentials of the very idea of such a process in the field of poetry. It is in three parts, somewhat loosely interrelated. The first part inquires into the legitimacy of inquiring into the "creative process"; the second describes some aspects of my own experience, to see whether anything in the processes of my creating deserves to be called a creative process; and the third asks why one should try to effect a union between such disparate concepts as those of creation and process.

I

It may seem reasonable that someone who has both published extensively on aesthetic theory and made a public profession of poetry should be asked to testify from personal experience on "the creative process." Yet a poet's first impulse when asked how he writes poems is usually to resist the question. "With pencil on paper," he will say; or "in English"; or "with difficulty"; and so on. These may be truths, but the questioner is likely to feel they are the wrong truths. Yet why should these not be the only truths there are? Why should there be a further question? Anything that ends in a poem must be a poetic process, and anything that ends in an original poem must be a creative process. What more could one say? A way of writing should have no interest for the public independent of what is written and published, and when the worth of what is written is established, it can be of no consequence whether it was written in this way or that. It is notorious that when poets talk among themselves they find little to say about the processes and procedures whereby they write. These have nothing interesting in common, and confessional anecdotes soon

D. Dutton and M. Krausz, eds., The Concept of Creativity in Science and Art, pp. 47-73.

grow tedious. The problems in which poets show an engrossing common interest are those of publicity and finance: how to get their work before a sufficient public at a rate that makes it not quite suicidal to devote some reasonable proportion of their lives to their exacting art. The process of writing enters into this concern only insofar as it is related to skill or luck in attracting commissions and similar opportunities. This has been said often; but, however often and emphatically it is said, it always needs to be said again.

Yet people persist in asking about "the creative process" as though that were something above and beyond writing poems or painting pictures. What makes them persist in the face of such discouragement? Perhaps there are two things that seem so puzzling that they feel they must return to their question. First, since poetry for most poets is not a living and never has been, and is less likely to bring fame than to incur rejections and contemptuous reviews, one wonders what drives anyone to choose it for a career. And second, since poems are seldom asked for and even less often fill any social or economic need, how does one decide on any particular occasion to write a poem, and how does one decide what poem one shall write?

Those do seem to be good questions. Yet to ask them pre-supposes one or both of two things, both strange or absurd. The first presupposition is that any activity that is not justified either by socio-economic yield or by an outcome successful in some other fashion is at best inexplicable and at worst suspect. But that is absurd, as Aristotle showed: some activities must be valued for themselves alone, or nothing can have value. The second supposition is that to write poetry is to do something odd or at least something requiring explanation: that poetry is somehow not normal. But that would be a strange thing to suppose. The occasional practice of poetry seems to be very widespread. A large proportion of our young people commit themselves to verse at least once in their lives, and regular poets are not especially rare; history fails to record any society in which poetry was not practiced, or in which the poet was not a recognized figure. Nor is it unusual, even in a society so dominated as ours is by its economic structure, for a person to devote time and energy to an uneconomic activity that interests him. So a poet asked about the poetic process may resent the question as expecting him to assume responsibility for an oddness and eccentricity that no tenable view of civilization would impute to him: the oddity, he may retort, is that civilization should have decayed to a point where a merely banausic viewpoint could claim such privilege.

The poet does not need to be so touchy. The presupposition might after all be not that the poet is a crank but that his role is of such singular importance that we are interested in how anyone comes to take it on and, having assumed it, fulfills its requirements. And there is after all a difficulty in principle in understanding how poets write poems. It is this difficulty that gives rise to the concept of a creative process and is reflected in the seemingly paradoxical nature of that concept. Poets and other artists engage in creative production on a regular basis. This regularity seems to show that there is a dependable or at least repeatable process or procedure they go through; otherwise, one could not say of a person that he was now a poet, but only that he had been one. To say that he is now a poet suggests that he will go on writing. Yet this seems something not even the poet could know, since our notion of a poem requires that it be original, hence unprecedented, hence unpredictable. Thus, to say that one is a poet is to predict the unpredictable.

Strictly, one does not know that one is still a poet, for every poet must one day write his last poem, and will not then know that it is his last poem he has written. But one may have good reason to believe that one is. And that good reason may not be that one has access to some "creative process," or some Muse who will continue to call on one when she is called on. Any one who knows he is a poet also has some knowledge of what sort of poet he is. His poems, however original, will be his poems and will manifest to himself as well as to his readers something of his characteristic style. The poet knows he is a poet because he has a way of writing, which he knows. When he stops writing it will most likely be because he is no longer interested in writing that way. If there is such a thing as a creative process it may lie in this, that every poet develops his own way of finding themes and his own way of working them. The process is to find a theme and work it.

That might be another good place to stop. Psychoanalysts and historians might find something more to say in general terms about how themes are found and worked. But the philosopher cannot: he can only insist that a theme must be found, and worked too. And the poet can only tread his own regress: somewhere at the end of his technique and at the bottom of his bag of tricks there must be an absolute starting point. But so it is with all skills, even the most rudimentary. How do you raise your arm? Philosophers have liked to ask that question, and conclude that though the arm-raiser may invent or discover means, procedures, or re-descriptions, sooner or later he must come to something that he admits he just *does*. Otherwise he would never get his arm up. One might postulate an "arm-raising process," just as one speaks of a "creative process," as a fancy way of saying that people just raise their arms, as poets just write poems, without being able to say altogether how. A poet is someone to whom writing a poem has become something as intimately familiar as raising his arm, the difference being that what he is intimately familiar with is the way of doing it that he has developed for himself.

In this more reasonable frame of mind, one can after all say something more general about the creative process. How does the poet think of something to write about? The answer must be: it just comes to him. And how does it come to him? Maybe somebody brings it. Bridegrooms evoke epithalamia as editors elicit articles. But if nobody brings anything, if there is no commission or request, how does it come to him? And now the inexperienced or very infrequent writer cannot say. It simply does. That is exactly what it is to be an inexperienced or infrequent writer, that there is no set condition on which he writes. But the experienced writer can say. He keeps on the watch for occasions, seeks them out. He scrutinizes his world for occasions of just such poems as he knows how to write. And how does he recognize such an occasion? Because he sees in it the possibility of just such development as lies within the scope of his practice. Again, that has to be exactly what it is to be an experienced writer. To have experience is nothing other than to be able to recognize and exploit occasions for skill. It might be thought that what this describes is the practice of the unoriginal and uncreative writer, but it is not so. Even the most astonishing innovator astonishes in the sum of his work, or the total development of his practice, or in one or two works that are new departures, and not in each work taken singly in relation to the others. On the other hand, the experience of the poet is not that of the farmer. The farmer must recognize the right day for grubbing the ruta-bagas, and it is the same day that another farmer would recognize, and would be an equally good day for anyone placed as he is placed to get them in. But what the poet has to recognize is the proper occasion for the exercise of his own style and no other, and it is rarely that even a sympathetic poet can suggest to another what such an occasion would be.

When the poet has thought of something to write about, how does he know what to write about it? The answer is almost as before: it comes to him. But not quite as before, because it must already have come to him. In recognizing an occasion or an opportunity, he must have recognized it as an occasion for doing this or that. But we may set that aside, for unless the completed work springs to his mind ready-made there is work yet to do. How does he know how to do this work? It comes to him. And how does it come? Again, it may be brought, or some of it may. It is possible that someone should tell him what metre to use, what rimes, what analogies, and so on to any extent. His ignorance may be thus aided, his expertness thus tested. If I give a child a paint-by-numbers set I suggest at once that he shall paint, what he shall paint, and how he shall paint it; and a less explicit variant on this procedure might stimulate production without quite excluding creation. But, of course, any part played by such intrusion is outside our interest when it is specifically the creative process that engages our attention.

Given a *donnée*, then, how does the knowledge of how to develop it come to the poet? Once more, the poet who lacks experience cannot say. He can know how to follow a rule, but not how to depart from it, and if he departs he must do it in fear and trembling or in foolhardiness. With the experienced poet it must be otherwise. Strictly, he cannot *say* how it comes to him, for anything he could say would amount to a rule he could cite. But, though he cannot describe it fully, there is nothing he knows better than his own way of proceeding. That, and nothing else, is what his art is. He knows what to do next, and this is not what the rule of his art prescribes but what determines his style. And the truly original and creative artist is the one whose style determines a way of developing and changing his ways of proceeding themselves, who sees in the next occasion for his art not an opportunity to do what he knows how to do but an opportunity to do what he knows to be the next thing.

But how does the poet know it will continue to come to him? The artist's perpetual fear that the sources of his inspiration will suddenly dry up seems real and reasonable, but there is something factitious about it, as though a man were to fear he might forget how to speak his native tongue — or how to raise his arm. Such things do happen. But in the ordinary course of events, as Aristotle remarks in opposing the suggestion that a man might "forget" his moral principles, one does not forget what one does all the time. I forget my French, which I speak seldom, but not my English, which I speak every hour and in which I frame all my thoughts.

Here is the heart of the matter. The artist, the poet, is to be known by his sustained habit of attention. As he scrutinizes his world for themes, so he looks ceaselessly up and down for ways of proceeding. His style is a style of search, not a habit of acceptance. His mind is a restless scanner, an inward rat. The other day an inventor was interviewed on television and asked if he might not run out of ideas. He said he would not. "I think all the time. If you think for forty hours a week, you'll think of something." The layman who asks the writer how he gets his ideas seems to think that such ideas would be forcing themselves spontaneously on a mind as idle as his own. But nothing is more evident to the artist than that he is working at his art, and the layman is not. What poets have most evidently in common is not a mysterious contact with secret springs, and certainly not any shared mental process, but simply a steady application to the actual writing of poetry. Anybody can be a poet who really wants to be, though wanting will not make him a good one. What the layman does not do, and probably could not do, is bring himself to attend steadily, day after day and year after year, to the business of the art.

To ask a poet to describe the creative process is to ask him to formulate a rule, or something that will do in place of a rule, by following which any idle ninny could make a poem. But writing poems is something idle ninnies cannot expect to do without forfeiting their idleness and ninnyhood. A poet is not an idle ninny who just happens to own a sort of magical sausage-machine that he might lend (or of which he might deliver the patent) to his neighbor, like lending him a power-mower. If there is a creative process it cannot be a substitute for intelligent work. It must be a way such work is done.

II

Talk of the "creative process," it seems, covers two questions the poet may be asked: how he came to bend his attention steadily in that unremunerative direction, and how he comes to write this or that poem. If I now answer these questions for my part, it is not because I wish to speak for "the poet," since each must answer for himself, but because what I think about these matters must reflect what has happened to me, and the way I interpret my experience must be colored by my theories. My answers will be honest, within limits, but I cannot answer for their truth. The way I now recall my life may not be at all the way it was.

In my assumption of the vatic mantle — a suitably portentous phrase, stressing the self-image rather than the work to be done; and a mantle, unlike a *persona*, covers the contours but leaves the face exposed — I recognize five stages and look to a possible sixth.

The first stage was my realization that the poetry we were taught at school was something I understood better than my teachers — or rather, understood in a way they did not. I had a feel for what was going on, like someone watching players at a game he knows. That was when I was a child, eleven or twelve. The second stage came a year or two years later. It was the realization that I knew how poetry worked, what governed the choice to say this rather than that. The poems I was then writing were not good ones, even by the standards appropriate to children's verse, but they were rooted in a confidence that I knew what I was doing — a confidence that I have never lost, and have never experienced in any other form of activity, even in those in which my practice seems objectively to have been more successful.

I have sharp and vivid recollections of the occasions on which the two realizations I have mentioned came to me — or rather, perhaps, of the moments that have come to stand for these discoveries. The difference between the two stages thus marked seems very real and evident to me, though I am less sure that I can put the sense of it into words: it is the difference between finding something congenial and finding it rational, between a feel for the whole and a grasp of its workings. But it may after all be that the real difference is between the two images I retain, and the two incidents they purport to stand for.

The third stage begins with the recognition, at fifteen or so, that poetry was what I came into the world to do: not something I *could* do, but something it was my business to do. R.G. Collingwood in his *Autobiography* tells how the conviction that philosophy was his business was precipitated in him at the age of eight, though he did not then know what the business was. I recognize in his account something akin to my own experience, and suspect it may be a common one in the onset of prophetic vocations.

The fourth stage is not linked to any date or event. It was the realization, somewhere in my twenties, that it was not in me to be a first-rate poet; specifically, that my gift was not such that it would be rational to organize my life around it or sacrifice all other pursuits and interests to it. I have called this a realization; but it may rather have been a decision, or even the entry on a new style of self-dramatization. But a decision or self-assessment was in any case called for. It is part of our romantic orthodoxy that devotion to any art must conflict (or demands readiness to conflict) with moral and social obligations; and even without that orthodoxy everyone must confront the question of what weight each of his major concerns is to have in the economy of his life. So I decided to be a minor poet. Poetry would be something interstitial, something I did when not preoccupied. But this was a change of policy, not of attitude; poetry still was, and still is, the only thing I took completely seriously. But I encased my seriousness with irony and formed my life around the inwardly farcical but outwardly respectable career of the academic philosopher.

The fifth stage began in 1958 when, at the prompting and with the help of a friend, I began to publish my work. From this point, poetry was no longer a private affair, a matter of my self-image, but part of my ordinary engagement in the world. Publication normalizes poetry.

A sixth stage would begin if I were to begin to write, or to believe myself to be able to write, extremely good poems. Poetry would then be something I would think it proper to give things up for. But this does not seem likely.

Such is my poetic process (wholesale). There seems to be a clear enough pattern to it: first seeing that a practice is congenial, then seeing it as a field for confident operation, then the sense that practice in this field is not merely possible but called for, and at the same time the recognition of the associated role and the decision to present oneself in one of its versions; and finally, in maturity, the recognition that the role is to be played in a certain fashion and under certain conditions suggested by an adult assessment of social reality and one's own place in it. But this clear pattern turns out not to be very interesting, because on inspection it proves to be one that everyone must follow who is to adopt a career on the basis of its attraction rather than because it is socially accessible. In fact, it seems likely that an analogous path must be traced even by those who merely drift along the line of least resistance into a hereditary niche. It too must seem at first merely congenial, because one has grown up with it; then feasible and intelligible; then, as one confronts adulthood, it must begin to figure as what one is going to do with one's life; then, as one matures in one's occupation, one comes to know one's place in it both as a set of skills and as a social position. And at last one simply occupies one's place, and no longer needs to think about that. So I could have spared my autobiography, had it not been that it was only by reflecting on it that I have come to recognize this form of development that the very nature of the case seems almost to demand.

But how about the poetic process (retail)? How do my poems get started? For anyone who makes a profession of poetry, there can be only one answer to this: in every way possible. In the practice of my profession I direct my attention upon all my surroundings in a hunt for occasions of writing. It must therefore be the case, and it is, that everything that could serve as such an occasion will actually serve as one. If I knew of any other sorts of occasion than those I find, I would make use of them at once. A poem may start from the following. An incident. A recollected incident. An incident read about. A mood, or more likely something seen and experienced through the mood, for a mood in itself is nothing. A phrase that suggests a verbal structure. A word that sticks in the mind, or a line or a snatch of rhythm that comes from nowhere. Some words chosen at random by sticking a pin in a dictionary — rather tricky, that. An idea for a kind of poem, or a scheme to be filled in. In fact, anything that a poem could be about, or that could go into a poem, or any way of writing a poem. Whatever. All that is necessary is that one should be able to say "I could use that"; one need not even know at once how one could use it.

Sometimes an idea for a poem, or a part for a poem, may occur to one when one is not on the watch and not in the mood for poetry. It may then stay around for years until eventually it germinates or is abandoned as after all boring or irrelevant. Sometimes a line, or a phrase or a rhythm, comes together with the adumbration of its own completion, a shadowy notion of a larger structure in which the fragment should occupy a place already determined. But in my case the adumbrated structures are usually large and complex, and I recall no case in which such a structure came to fruition: the pattern completed has always been a smaller one. No doubt that is part of what it is to be a minor poet, that the settled tendency of one's work is towards a diminution of scale.

One possible source of ideas I have seldom used: dreams often yield poetic fragments, but these seldom lead beyond themselves or sustain their promise of interest. And I have chosen not to resort to drugs for the release of inhibition or the derangement of the senses.

Of a very few poems I cannot say how they originated because they appeared spontaneously completed, requiring only to be written down and slightly revised. These poems do not seem on the whole to be better or worse than those in whose genesis deliberation played a greater part, and differ from them in only two ways: that they contain rather more banal "Freudian" imagery (of pegs and holes, birds and oceans), and that they tend to contain weak elements that for a long time escape notice or resist revision. These spontaneous poems seem neither more nor less characteristic of me than the others. And indeed one does not see why any great gulf should be fixed between the willed and the unwilled. Someone who habitually writes verse might be expected to develop such a facility that he could and sometimes would do so without thinking about it, much as an experienced driver can direct his car through heavy traffic without giving any thought to where he is going.

It is often thought that the operations of the unconscious mind in composition (to which the "instantaneous" or "spontaneous" poetizing I have referred to might be thought to attest) are accompanied by agonies of "incubation" and other psychological disturbances, and that these are an integral part of the "creative process." My own experience has been that such pains, irritabilities, abstractions, fatigues, and other signs of unconscious effort are not particularly associated with poetry or creative writing. They occur at a certain phase in the composition of philosophical writings, or the solution of personal or administrative problems, or any complex matter. As many have reported, the phase is that at which a problem is mooted but the terms in which it is formulated are too nebulous, or too complex, to repay systematic problem-solving, any attempt at systematic or step-by-step procedures being at once frustrated by the conviction that the terms of the problem are unclear or just wrong. It is then that one waits and agonizes, contracts a migraine, quarrels with one's family, goes for long walks, dreams, or tosses in insomnia. These distressing episodes seem in my own experience to be related to the phase in the problem to be solved, and to have no connection with antecedent or current events in my personal life. The popular image of "the oyster and the pearl," according to which such distresses bear witness to the nacreous covering of an intrusive irritant in one's psyche, seems based on an interpretation of only one class of cases — and, I suspect, a conjectural interpretation at that. The painful incubation marks rather the phase in problem-solving that calls for intensive work of a kind one cannot consciously do.

My reference to unconscious labors is open to the objection that, even if the very notion of an unconscious mental activity does not embody a contradiction, unconscious activities are *ex hypothesi* unobservable and hence merely hypothetical constructs: all that is observably happening is that one is interrupting one's work with a bad conscience. The objection is easy and has point, but it is misleading to speak thus of "all that is observably happening," for that is not sufficiently described unless one says that everything goes on as if an internal energy-consumer and attention-distracter were scanning one's problem for a system of ordering that would allow the work of conscious problem-solving and monitoring to go forward fruitfully in a systematic or serial fashion. Unconscious mental activity is a dispensable or an empty hypothesis only in the sense that the physical world is so.

In addition to the occasions for verse mentioned, there are commissions — requests for a piece of writing of a certain sort, for a certain occasion, of a certain dimension, on a given topic — any or all of these. In this regard a poem is no different from a philosophical paper or lecture. I can accept such a commission in full confidence that I will come up with something on schedule that will fulfill the requirements, although of course I cannot answer for the quality beyond a certain minimal competence. Why should I not be confident? It is my business to do so.[1] Between the acceptance of the commission and the fulfillment I shall expect the pains of incubation to intervene, but sometimes they do not; sometimes I am visited by what seems to me an "inspiration" in the form of a genuinely novel and exciting idea and sometimes I am not; and the occurrence of episodes of either kind bears no observable relation to the degree or kind of satisfaction the relevant public takes in the outcome.

In outlining my career as a poet and cataloguing the occasions of my poems, I have answered both the questions that inquiry into the creative process seemed to comprise. But I have left out what is superficially the most striking aspect of my experience, which in this respect parallels that of another academic part-time poet, A.E. Housman. Aside from my adolescence, when one writes verse as a baby dribbles, there have been four periods in my life when I wrote verse regularly, a poem or a large part of one more or less everyday. In Housman's case, these periods were by his own account times of excitement, answering no doubt to gusts of inspirational afflatus or surges of hormonal flow. In my case, no such disturbance is easy to discern, though of course there is no knowing what an analyst or a lie-detector might dredge up.

The first of these periods of regular production, from January 1 to the middle of March, 1944, was undertaken as a matter of policy in the context of adoption of the role of verse-writer, the date fixed in advance — one begins diaries with the new year. So any relation to personal disturbance must at least have been mediated.

The second period, in January, 1966, had a double occasion that again called for no troubling of the waters by angels. In the first place, a hiatus between teaching engagements had left my mind free from preoccupations for the first time in years; in the second place, recent publication of a volume had depleted my stock of unpublished poems. Again, the decision to go into regular production on a certain date and for a certain time was taken beforehand. On this occasion I confirmed that I could, given a clear mind, sit down with a sheet of paper and be sure that in no more than an hour I would certainly have come up with a poem or a substantial part of one; though that does not mean that I did not, or need not, keep on the watch for the rest of the day.

The third period of regular composition, from late August to October, 1969, was different. A reading of Basshō's *Journey to the Deep North* reactivated an old interest in Japanese verse and occasioned the project of writing a series of occasional poems in Japanese forms, at least one every day. These, being very short, could be composed in short times of peace, without needing protracted freedom from preoccupations. Not surprisingly in view of the nature of the initial stimulus, the series of poems that resulted tended to be on the theme of travel, but in fact it turned out to be more cohesive than that and ended in a sequence that took on the evident though unheralded character of an end. One might therefore postulate some internal unconscious dynamic governing the form of the whole, that some form of structural or psychological analysis might lay bare. However, the poems did not, as a sequence or individually, come from any felt disturbance or other specific experience other than that of reading a book that suggested a model.

The fourth period, from June to October, 1970, presents a more complex case. Superficially, the situation was as in 1966. The recent publication of a book had depleted my unpublished store, and the start of a sabbatical year provided the necessary freedom from constraining preoccupations. But although I might, conformably with past practice, have provided for a period of regular poetizing, I made no such plans. Rather, the book was occasioned by a curious incident. I had a dream in which I arrived at a meeting where I was to read from my poems, and found I had brought none with me. I hurried home and searched in vain for my manuscripts, but at last found in a drawer a book I had forgotten. This proved to be full of poems I did not recall writing but knew I had written; they were full of a strange power. On waking, I wrote down what I had dreamed — something I do not usually do. Some time later, I found in my office a mimeographed document that I recognized at once as the book of my dream, and which I knew at once that it was laid upon me to fill with poems. So I did so, one or more on every day when I was not sick or travelling. This time, though, there was no conclusion. It simply happened that by late September there were fewer and fewer days when I could summon the composure to wait in stillness. The initial sense of something akin to obligation had given way to a mere habit of continuing, and after a while there seemed no reason to push it further; besides, I was becoming more engrossed in a complex piece of philosophizing. So the series dwindled away, its last members perfunctory and fragmentary, and the book remains unfilled. The poems, of course, do not have the special power of the poems I dreamed, and differ from other things I have written only in the absence of any large and complex structures. The reason I call this incident curious is that next winter I found the account of the dream I had written, and found that the book I had dreamed was not really the book I "recognized" — it was in fact a book I knew well, the account book I keep for the income-tax people. It is as if the dream and the false recognition were part of some unconscious dynamic in me, though the outcome fitted well into what would have been a normal and sensible strategy for me to pursue deliberately. And I note in myself an unreasoned urge to tell the story, in one form or another — in fact, it is only to provide a setting for it that I accepted an editor's suggestion that I might exploit my own experience for this article.

So much for anecdote. I infer from my experience that there may well be an unconscious dynamic of composition, but that it is useless to look to it for any characteristic significance, idle to expect that it will correlate with any character discernible in the outcome, and rash to suppose that it always works in a way that is independent of the conscious strategies of its host.

A caution is needed. Obviously I recall much in my life that I have left out, although it could affect the interpretation of what I have written. For instance, the occurrence of sustained but widely separated bursts of poetic activity might seem odder than it is, because my narrative did not mention that my writing of philosophical prose has accustomed me to regular writing for protracted periods on set themes. But so it is with all those anecdotes of sudden inspirations and dream-compositions and emotional disturbances that figure so often in writings on the creative process. Such accounts necessarily come to us as parts of stories that are the products of selection and editing; and it often happens that the selecting and editing are carried out in order to present a certain image of creative procedure, and may therefore omit material that could have been used to tell a very different story — one, perhaps, in which the striking incidents would not figure at all, or would not appear striking.

III

The phrase "the creative process" is used in two different ways. Sometimes it is used generically to refer to all processes, whatever they may be, whose outcome meets some appropriate criteria for originality. At other times it is used to refer to some specific process whose characteristic outcome is supposed to be original work. It is not always clear in which sense the phrase is being used, and the distinction is not always recognized. Yet the distinction must be made, for if original work is ever produced it must be produced somehow or other and the phrase cannot lack application in the former sense; but it does not follow that the phrase has any application in its latter sense.

Three different accounts of "the creative process" have achieved some currency. One, associated with Paul Valéry,[2] and ultimately inspired by Edgar Allan Poe,[3] is exemplified in Part II of this paper: it reduces to saying that some possible component or aspect of a poem serves as a starting point and is then elaborated, partly heuristically and partly systematically. Since all this says is that one must start and then continue, this is clearly taking "creative process" in the former of the senses distinguished above. It tells nothing about how poems are made, except that they are poems and they are made. A second account of "the creative process" is associated with Robert Graves,[4] T.S. Eliot,[5] and a host of romantic writers.[6] A schematized version of it might go as follows. First comes the original formulation of a problem or conceiving of a theme (preceded possibly by the poet finding himself in an excited, troubled, or sensitized condition). Second comes a period of random search not directed by the will, unconscious incubation, and so on. Third comes a flash of insight, a relief from suspenseful tension, a sense of how things will go together. Fourth comes the deliberate elaboration of this insight; and fifth is the criticism and refinement of the elaborated solution. This five-stage model will be referred to from now on as "the standard version," and seems to take "creative process" in its second sense, for it appears to describe a psychological process whose correlation with creative solutions to problems or original achievements would need to be established empirically.[7] The third account of the "creative process" to achieve currency is that elaborated by Arthur Koestler from a hint of Freud's.[8] Here, creativity is attributed to the unforeseen interaction of two or more thought-patterns previously elaborated independently. But does this take "creative process" in the former or the latter of the senses we distinguished? It might be either. Although it seems merely to say how originality actually comes about in typical cases, it might be construed to mean that in every truly original work one must be able to find a complexity of the kind indicated, so that we are merely performing a logical analysis that must in some form or other be presumed to have its counterpart in the process of production. The reason it is hard to tell which is meant is that many authors who write in this field think in terms of a natural necessity: the mind is a logical engine, so that in describing what takes place we are at the same time performing a logical analysis, and vice versa. I return to this question later.

In whichever sense the phrase "the creative process" is taken, and whichever of the three models we adopt, we may be met by the objection that the very phrase embodies a *contradictio in adjecto*. The idea of creation is that of production from nothing, production of absolute novelty, and the absoluteness of the novelty requires that there be no rule, no method, no series of stages, and accordingly no process whereby it came. The model of creation is the unqualified fiat of God, applied to names whose sense is eternally or instantly complete and only awaits its reference. But the model of a process is something like digestion, in which one can identify initial and terminal stages and enumerate and explain the steps whereby beginning is transformed into end.

To see that the concept of a creative process is not so paradoxical after all, one might compare Aristotle's seemingly innocuous account of the requirements of a whole.[9] It must, he says, have a beginning, a middle, and an end: the beginning is what has no necessary antecedent but has a necessary or normal consequent, the end is what comes naturally − necessarily or normally − from an antecedent but has no consequent, and the middle of course has both antecedent and consequent. So we have a process divisible into linked stages; but the accounts of beginning and end are not symmetrical. The end is not said to necessitate its antecedents: it is not a ground of explanation, but something explained. The beginning, by contrast, is a ground of explanation but is unexplained. The concept of a beginning thus defined is that of a novelty that is absolute but fertile.[10] But if the beginning is an unexampled beginning, the necessary steps by which the inevitable end is reached cannot be known in advance. It must be a necessity retrospectively revealed. And if we recall the way Aristotle usually speaks of natural necessities we may suggest that what is retrospectively revealed is how the intermediate steps were needed to bring about the inevitability of the result. So we may say that a creative process differs from a non-creative process in that the causal connections are not known in advance, as they are in processes where the initial stage is itself explicable as how what regularly happens normally begins. One might still complain that the notion of a process is misapplied, as though anatomizing a whole were tracing the steps of its genesis; but the charge of paradox will not stand. In fact, recent philosophical discussions of creativity take two forms. One inquires into how something akin to rules and methods, right and wrong ways of proceeding, can enter into the processes of producing unprecedented objects. The other inquiries into the nature and locus of the novelty in the created work, without supposing anything about the way it came into being. The former is no more open to the charge of paradox than the latter.

Let us consider further our Aristotelian model of a creative process as one with an absolute beginning and a series of subsequent stages whose necessity is therefore only retrospectively revealed. A poem or other work of art obviously functions as some sort of communication among people, and one might therefore expect to reduce it to terms of the simplest model of communication theory. This model assumes that one starts with something determinate to be communicated. This is encoded, the encoded message is transmitted, then at the other end it is decoded and the original message recovered. The model calls for no absolute beginning: the original message is a structured datum that enters into the communicative process for reasons congruent with the formulation of the message and the devising of the process itself. Let us contrast with this R.G. Collingwood's account of artistic expression, which we may construe as a standard philosophical account of the creative process.[11] Here there is no determinate first stage, no known and structured message to be encoded. There is only an encoding that is at the same time a deciphering. The creative process is a passage from unclarity to clarity, the imparting of structure to inchoate feeling. But this is seen not as a passage from beginning to end, as though the clarification were brought about by the successive application of ever finer filters, but as the arrival at an absolute beginning. So there is and there is not a process: there is, because the artist has progressed from unclarity to clarity, and what he has clarified is in an unexplained sense the "same feeling" that he began with;[12] but there is not, because the end has the status of an absolute beginning in the light of which earlier phases are of no account, sublated, *aufgehoben,* or something. And what is transmitted from artist to public is supposed not to be in a code or to admit of decoding, because the artist's encoding was also a deciphering and all paths of transmission are magically bypassed. This is strange stuff indeed, but the strangeness may be demanded by the nature of the case. Attempts to apply information theory to art seems to fail just because they see no need for such oddness.[13] They simply take it for granted that a work of art is the encoded version of a pre-existing message that the artist wishes to transmit. It is only on this supposition that the concept of redundancy can be pressed into service to give precision to the notion of style. But if in fact there was no original message — for otherwise the process would have been reproductive rather than creative — there is no way it is transmitted, no code, no bits of information, no redundancy. The model cannot be applied at all.

Where there is no initial message to be encoded and transmitted, we cannot tell message from noise. The creative process is thus necessarily indeterminate. Yet it must be strongly ordered, since the end is ordered. Vincent Tomas has spelled out how this is possible, in an account that we could crudely adapt to the terms of Part I of this paper as follows.[14] The work must start from a *donnée* that has to be regarded as gratuitous, an absolute starting point. But the artist perceives it as structured in a way that permits and calls for a line of further development. It is fidelity to this line of development that serves to regulate his continuing creation. Normally, it is the author's style, which includes his style of changing his style, that serves as a matrix both for the structures originally discerned and for the continuations suggested, though of course it may happen that the development envisaged is in some ways, even in many ways, unprecedented. In any case what serves as controlling factor is the present condition of the uncompleted work in the light of the possibilities originally contemplated. But we have to add that every recognizable intermediate stage in the work's progress serves as a new *donnée* suggesting a partially new set of possibilities for development. The work is presumed complete when a stage is reached that calls for no development.

This modest account of something that is both truly creative and truly a process, controlled in a way that can be rationally accounted for and retrospectively described but is at the same time unprecedented and free from rules and routines, is meant to exclude an account in Collingwood's manner that supposes the end of a work to be somehow implicit in its beginning. To see a way of proceeding is not to sense the way to a goal. True, but one feels that the sense of an end must be operative in some way or other. The artist's activity is inexplicable unless he has a hunch that a completed work might result from his working; and his sense of work to be done and progress to be made is unintelligible without some adumbration of a sort of completion. Accounts of the sort I have outlined should be taken not as denying that such a sense of an end is necessary but as denying that it can be understood as determining the course a work takes. The envisaged end is likely to change as the work already done reveals an unsuspected character and suggests unforeseen possibilities.[15]

It seems that one can produce at least one account of what a creative process would be that is coherent and free of paradox. But we have not described a definite process. All we have done is show that the notion of creation does not exclude everything that might be termed a process, and suggest how such a process might be articulated. This is not to say that the notion of creativity includes or implies any sort of process. To speak of creation or creativity is not to say that a sort of procedure has been followed, but only that certain sorts of procedures, namely routines, have not been followed. What, for instance, do we mean when we call a person "creative"? Sometimes we mean that his solutions to problems are unexpected and fruitful — unexpected, or they would be routine; fruitful, or nothing would have been brought into being. Sometimes we are merely classifying his job as one that involves neither routine production nor dealing with the public: he works in the creative department, no matter what the merits of his work. Sometimes I think we call a person creative if his rate of production in the fine arts or some other creative activity is unusually high: if he is fertile in fruitful solutions, or if he is full of ideas. In general, a person is called creative because of a tendency to produce things of a novel sort, or things of a sort in which novelty is important, in a variety of contexts. In no case do we suggest anything about his procedures of production. One might go further and insist that it is absurd to suppose that a single process would characterize such a tendency in all fields alike. Creative activity is not a special sort of activity, but any sort of activity that issues in new being. The idea of a creative process is not paradoxical, but gratuitous.

Those who think of "the creative process" as a characteristic sort of sequence of psychological events are exposed by the foregoing considerations to a familiar philosophical attack that seems crushing.[16] The occurrence of any such process, it is said, is logically incapable of being either a necessary or sufficient condition of creativity in the sense sketched above. Novelty and fruitfulness in solutions can be identified without knowing their psychological antecedents, and a hackneyed or eccentric solution is not redeemed by any anecdote about its origins. But this argument, though valid and important, does not suffice to show that the occurrence of a certain sort of process in a context where the novelty of the outcome is an important issue has merely anecdotal significance. It shows only that the process cannot serve as hallmark for the product, as some writers have seemed to think it could. But that is not the only significance it could have.

Suppose someone were to put forward our five-stage "standard version" as a characteristic way of producing novel solutions, in the arts or elsewhere. This might indeed be put forward merely as a generalization about how originators in fact originate. Even that would not be merely nugatory: if we are interested in a result, it is natural that we should be interested in how it came about, and the way things tend to happen in the human world is a proper matter for concern. Alternatively, one might propose it, not as necessary or sufficient condition of creativity already recognized, but as actually defining a sense of "originality" — that sense, in fact, which is relevant to art. In some cultures, a pot is not deemed properly made unless its design was revealed in a dream.[17] To do this is of course to propose a radical revision of our ideas. Art is transformed into something of which a psychological process could be the necessary and sufficient condition — that is, a psychological episode. The work of art is reduced to the status of evidence that "art has taken place."[18] All criticism is reduced to foolishness because the evident object is replaced by an unverifiable claim. And, because the alleged dynamic is common to creative work in any field and not in the arts alone, it replaces the role of "artist" in our hagiography by that of the creative person conceived as a psychological type, a sort of holy fool. Such a revision seems to me preposterous, but a few years ago there were many to propose it.

These extreme interpretations of our "standard version," as anecdotal and as definitive, are not the only ones possible. One could for instance argue that the occurrence of the process, because its very description seems to testify both to a serious churning and to some sort of resolution in the outcome, constitutes a claim that the outcome merits attention. This might not be a claim of actual worth, for one might well concede that puddings are proved in eatings, but a claim that what has emerged merits scrutiny for possible worth. This point may be obscured by an inevitable but misleading tendency in collections of reports on the creative process. Only successful artists are asked to provide anecdotes of how they create, so that what they say may appear to be a recipe for success even though a thousand bunglers might have told precisely similar stories if anyone had thought to ask them. Similarly, the handful of exemplary anecdotes that are endlessly repeated from the annals of scientific discovery report only those dreams and inspirations whose outcome was a successful discovery or invention. We are not told of the inspiration that proved abortive, the dream that dissolved on waking, the unconscious incubation and flash of insight whose elaboration disclosed an egregious blunder, the "got it" phenomenon whose sequel was a "lost it" experience. So, if we attach any weight to the occurrence of any psychological process, it would be wiser to do so on the basis of our general expectations about human affairs than on the basis of analogies from success stories.

There is a better reason for invoking the creative process as a clue to seriousness in the arts than in the sciences. Notoriously, the work of a truly original artist may be misprized until familiarity has revealed the right way to look at his work, the perspective from which his patterns are visible. We are therefore tempted to look to something other than what we find in the work for testimony to its seriousness or even to its worth. This extraneous aid could be the testimony of an informed critic, or the artist's proved reputation won by work in a more familiar mode. But it could also be something extraneous in the production of his work, such as his labor and eventual satisfaction; or, perhaps rather less extraneous, the occurrence of a process that we deem likely on the face of it to issue in serious work. We may think of the creative process as guaranteeing at least some sort of novelty, and, if the novelty for which it is a recipe is by no means the originality of the highest form of art, our description of it may serve as a guarantee that in a serious and experienced worker the outcome can be neither hackwork nor a mere flash in the brain pan. Even so qualified an assurance may be of comfort to the humble camp-follower of the arts in these difficult days.

If occurrence of the sort of process sketched in our "standard version" can be taken as presumptive evidence of seriousness, part of the reason may be that what the version describes is what we think ought to take place when something truly original is done. It may be that the standard version and its variants are popular, not because there is any reason to think that anything of the sort often or ever takes place, but because it seems fitting: a work of art is as if it were the outcome of such a process. As human beings, we feel able to pronounce on the fittingness and congruence of human affairs and attitudes. We do not feel the need to have recourse to the behavioral sciences to assure us that our sense of how life goes has empirical backing. On the contrary, we feel free to deride these sciences if they either otiosely confirm or absurdly conflict with our lifelong experience of being human.

It seems then that our standard version may function less as an inductive generalization than as an expression of our untutored and uncriticized sense of what ought to be the case, our way of imagining what it must be to create. It is but one step more to claim for it the status of a myth. Plato in his *Symposium* metaphorically describes artistic and intellectual creation as "giving birth in beauty."[19] Our standard account may be construed as an explanation of this metaphor. The triggering experience answers to insemination, the initial formulation of the problem corresponds to conception. Unconscious incubation is transparently an image of gestation, and the moment of intuition is the moment of birth. Then nature yields to nurture: the phases of elaboration and criticism answer respectively to the phases of development in which the infant masters first bodily movement and speech, and then social skills. Our reasons for adopting this natal metaphor are doubtless as much religious as evidential. It testifies to our sense of the dignity of art, to the importance we assign to the birth of new being, to the status of the work of art or the scientific discovery as a new and almost living force come into the world. It is a late survival of the eighteenth-century notion of the "genius" as someone who is less a fabricator than the place where an autonomous and natural evolution occurs.[20]

Our reasons for accepting this or that description of a creative process as what inevitably, or usually, or typically, or properly, or occasionally but interestingly takes place, may be of many kinds. But, as the examples of Tomas and Koestler may suggest, we might arrive at the notion of such a process simply by considering the conditions without which the emergence of something that was both unexampled and significant would not be possible. As a loaf must be baked, a poem must be created, and it is not clear why there can be no necessary conditions for either process. But at this point we would do well to abandon psychology for phenomenology or conceptual analysis. The alleged "process" becomes an elucidation of origination as such. If there is to be origination there must be a starting point of some kind; if the originated whole is to be original the principles of its organization must themselves be unprecedented, and must therefore represent a second origin; if there is organization, there must be elaboration; and if there is to be a public outcome it must have been promulgated and must therefore have passed the critical scrutiny of the creator or his custodian. Our standard version does little more than spell out these requirements in psychological terms, assigning each phase a place in a temporal sequence. The one item in the standard version quite unaccounted for is the "unconscious incubation," which from the present point of view is either a gratuitous piece of psychologizing (or theologizing)[21] embroidery, or else a dramatization of our recognition that the two starting points are in principle separate and that each is a true beginning. The transposition of the creative process from psychological and genetical to analytical terms is exemplified in a vigorous argument by Monroe C. Beardsley, who urges that what we call creativity and originality is properly to be located within the work of art itself, and is the emergence of complex resultant qualities that could not have been inferred from an enumeration of the constituents and their relations: this non-inferrability is popularly construed as unpredictability, put into a temporal frame of reference, and projected back onto the "artist" — as though works of art were produced by human beings instead of appearing, as we all know, ready made on the walls of galleries.[22]

Although proponents of the old "new" criticism may be contented with an account of creativity and originality in analytical or phenomenological terms, there are reasons why artists and cultural critics cannot be. There is an important difference between human and divine creation: the Creator's claim to originality is (in orthodox theology) unimpeachable, but human beings must establish their bona fides. Creation, properly, is the production of new being, of something that did not before exist but now has a being of its own and in its own right. The concept of creation is then basically less that of process and outcome than that of an achievement. What matters is that there should now be something that owes its being to another, God or man, but in itself is perfectly real. In the paradigm case of the divine creation, this suffices: before, there was no world, but now there is a world, and God will answer for it. But when we speak of human achievements an ambiguity creeps in. With half our minds we think that achievement, creation and originality are relative to their point of origin. We do not deny originality or creativity to either Darwin or Wallace on the other's account, and there is a sense in which Darwin's achievement would have been no less if Wallace had reached his conclusions a decade before but left them unpublished. So it is that to sustain a plagiarism suit the plaintiff must show not only that the form produced by the alleged plagiarist coincides with the pre-existing form, but that he had access to it. It may be in part because memory is delusive, the external history of creation often unknown and unverifiable, and the channels whereby ideas are transmitted manifold and hard to trace, that we like to invoke a creative process, an agonizing labor that at least shows that the alleged creator did not know he was doing anything so undemanding as plagiarizing. Just so, councillors were at one time summoned to attend the birth of a prince, lest some low-born brat be smuggled in in a warming-pan; but, if decorum precludes observation of the royal childbed, some reassurance may be had from the observation that the Queen was previously pregnant and is pregnant no more, and that the putative birth was heralded by appropriate groans.

It may be that the status of our standard version is more that of a parable than of a myth. The hard-headed will say that when the Pueblo potter says she dreamed her design the real purport of her claim is that she did not copy it. So it may be that to claim to have undergone the creative process is but a dramatic way of claiming true authorship. And why claim true authorship? Copyrights may be involved, and the artist must live. But that is not all. The value of originality has become deeply embedded in our whole way of thinking about art and even about science.[23] We think of art in terms of its history. The truest work of art, perhaps the only true work of art, we think, is one fit to figure in the history of art: the only true point is a turning point.

Victoria College
University of Toronto

Notes

*Reprinted from *Philosophy and Literature* 1 (1977). © The University of Michigan — Dearborn

1. It is my good fortune that my productive processes are properly synchronized with anticipated deadlines. Many unfortunate artists and writers are so constituted that the passing of a deadline is needed to initiate the process, just as some people are unable to set out for the opera before curtain-time.
2. E.g., Paul Valéry, *Aesthetics* (New York: Pantheon, 1964), p. 130 ff.
3. E.A. Poe, "The Philosophy of Composition," *Collected Works* (New York: Crowell, 1902), vol. 14.
4. Robert Graves, *The Crowning Privilege* (Harmondsworth: Penguin Books, 1959), p. 214.
5. T.S. Eliot, "Tradition and the Individual Talent," in *The Sacred Wood* (London: Methuen, 1920); and other essays.
6. Cf. F.E. Sparshott, "Xanthippe," in *Looking for Philosophy* (Montreal: McGill-Queen's University Press, 1972).
7. At a recent general meeting of the League of Canadian Poets at Fredericton, N.B., the three contributors to a panel on the writing of a poem gave accounts all of which conformed to the following pattern. The poet is attracted by a large public or personal theme which seems to call for a poem, but sees no way of setting about it. Then, an incident in the poet's life reminds of him of an incident connected with the troublesome theme, and in a state of excitement he writes a shortish poem or fragment interpreting one in terms of the other. On the basis of this he is then able to work systematically at the large theme. The pattern itself seems as interesting as the coincidence: this might be a typical way of finding an authentic mode of entry into a major theme.
8. See Arthur Koestler, *The Act of Creation* (New York: Macmillan, 1964), as well as the article by Koestler contained in this volume.

9. Aristotle, *Poetics* vii, 1450b26—31.
10. The notion of an absolute end brings an analogous inexplicability. How can anything have no consequences? Only in the heterocosms, the separate universes, of art, can an end be a completion that is not also the beginning of something.
11. See R.G. Collingwood, *The Principles of Art* (Oxford: Clarendon Press, 1938).
12. See, however, Collingwood's *Essay on Philosophical Method* (Oxford: Clarendon Press, 1933) for an explication of the patterns of thought involved.
13. Cf. A.A. Moles, *Information Theory and Esthetic Perception* (Urbana: University of Illinois Press, 1966).
14. Vincent Tomas, "Creativity in Art," *Philosophical Review* 67 (1958): 1—15.
15. Conformably with this recognition, Mikel Dufrenne, who speaks of the artist as responding to the call of the as yet uncreated work, describes it as an *indeterminate* call. See Mikel Dufrenne, *Phenomenology of Aesthetic Experience* (Evanston: Northwestern University Press, 1973), p. 35.
16. For example, see John Hospers, "The Concept of Artistic Expression," in *Proceedings of the Aristotelian Society*, 1954—55.
17. Ruth L. Bunzel, *The Pueblo Potter* (New York: Columbia University Press, 1929).
18. Here is an example plucked from the remainder table: "It could be argued . . . that the painting was a kind of relic, a kind of certificate or guarantee that certain activities had taken place previously which you were not there to witness." Donald Carroll and Edward Lucie-Smith, *Movements in Modern Art* (New York: Horizon, 1973), p. 132.
19. Plato, *Symposium* 206b—212a.
20. See M.H. Abrams, *The Mirror and the Lamp* (New York: Oxford University Press, 1953), chap. 8.
21. The phrase "unconscious incubation" seems to have had its original home in William James's *The Varieties of Religious Experience* (London: Longman's, Green and Co., 1902), with the phenomena of religious conversion.
22. Monroe C. Beardsley, "On the Creation of Art," *Journal of Aesthetics and Art Criticism* 23 (1965): 291—305.
23. For an early manifestation cf. Edward Young, *Conjectures on Original Composition*, 1759.

CRITERIA OF CREATIVITY*

CARL R. HAUSMAN

In *The Courage to Create,*[1] Rollo May defines creativity, in its "authentic" form, as "the process of *bringing something new into being.*"[2] In the course of his discussion, he considers several kinds of experiences which he claims to be essential moments of the creative process: "encounter," "courage," and a "drive toward form," among others. Beyond this, there is little attempt to explore the presuppositions and the complexities of the concept of *"bringing something new into being."*

I do not cite May's book in order to criticize it or to minimize its potential value for the layman and perhaps for some psychologists and psychiatrists. Without question, the book offers an interesting perspective on some of the problems connected with the creative process. However, as in so much of the literature on the subject, the book displays a conspicuous lack of attention to the task of providing a conceptual framework within which creativity is to be discussed.

The purpose of the following remarks is to suggest what I believe are the most obvious, basic expectations without which neither an activity nor its outcome would be regarded as an example of creativity. My account of these expectations attempts to specify criteria or necessary conditions of creativity based upon a characterization of newness. These criteria should serve as the fundamental limits of a conceptual framework which must be presupposed in a discussion of creative acts. But let me emphasize at the outset that I do not intend to write "creatively" about creativity. Nor do I intend to rest my case on a series of examples which are assumed to be instances of creativity. To the contrary, what is needed at the beginning of a study of creativity is a sober, straightforward structural account — an account that avoids as much as possible metaphorically couched expressions of praise for what is creative, and which postpones claims to provide immediate "insight" into

D. Dutton and M. Krausz, eds., The Concept of Creativity in Science and Art, pp. 75-89.

the nature of creativity. Thus, in what follows I shall attempt to lay out a bare conceptual skeleton, leaving for other discussions the attempt to fill out the flesh for the skeleton.[3] My approach is phenomenological in its basic orientation, but phenomenological primarily in the sense that I intend to focus on the most general, necessary ingredients in creative achievement.

An attempt to find criteria can be undertaken in one or more of three possible ways. We can proceed inductively with no preconceptions about which things are viewed as examples of creativity. Thus, we can simply begin with an enumeration and a description of processes that have been called "creative" and outcomes that have been called "creations." Or we can begin by stipulating which instances are to be included and which excluded. Finally, we can combine these approaches. If we were to proceed inductively and rely on the way the term "creativity" has occurred in all its uses, we would have to include such examples as home-making, salesmanship, and beautiful scenes in nature, as well as the work of Giotto, Shakespeare, Beethoven, and Einstein. This approach, I think, includes too much and spreads the reference of "creativity" too thin. On the other hand, if we stipulate the exclusion of such things as creative salesmanship, nature, etc., the approach, unless justified, is open to the charge of arbitrariness. The third approach, which I have in fact adopted, is to proceed descriptively under the constraints of the premise that there is a select range of phenomena which are most clearly and unquestionably examples of creative acts and of outcomes that are creations. The examples of creativity which will serve as a basis for the criteria are selected from the most dramatic and renowned instances of those outcomes that have been called "creations."

It will be noticed that I have referred here to the outcome of creative acts. The term "outcome" is not used to refer to results or effects of necessary and sufficient conditions. This usage would belie my view that creativity is not fully amenable to rational explanation. "Outcome" here refers only to the terminus of an activity, to whatever an activity articulates and to what is publicly accessible. Attention is directed to such outcomes because I see no way to distinguish one act from another without reference to what the act manifests — to something discriminable in the form in which the act reaches completion. And the form in which an act reaches completion can be found only in the outcome.

The most striking or eminent outcomes are chosen because these should show most clearly and sharply what called creativity to our attention. It is these which have been and are exemplary for other less dramatic outcomes, such as the results of cooking, or gardening, that are sometimes called creations. If an inquirer wishes to find some degree of creativity in hobbies and do-it-yourself projects — in everyone's activity — then what he will look for as creative in these activities has been drawn from encounters with dramatic cases.

I

The criteria of creativity which I shall consider are telescoped in the following claim: creativity occurs on condition that a new and valuable intelligibility comes into being. We begin, then, with the basis for three criteria: newness, newness of intelligibility, and newness of valuable intelligibility. Discussion of the three key terms, "new," "valuable," and "intelligibility" will serve as an account of these criteria. I shall begin with the term "new." But it will be seen that examination of this term cannot be understood apart from the others. As criteria of creativity, these terms are interdependent.

The outcome of an act which is considered creative, then, must seem to be new or novel. We might use other words to suggest this, such as "original," "fresh," or "innovative." These, however, depend on the acknowledgment of the fundamental and more common characteristic, newness or novelty. But what is newness?

A minimal condition or common denominator of newness is that it is present where something is different from its past. A thing that is new must at least be different from what preceded it. In this sense, every discriminable item is new, as a singular. It can be regarded as unique, as a discrete, specific item which is identifiably distinct with respect to all other items. It may be unique as a single instance of general, repeatable types or kinds. Each blade of grass, grain of sand, and molecule is different from every other blade of grass, grain of sand, and molecule. Likewise, what is unique may be a general or universal. Angularity, the quality of hardness, or a law of nature — each has an identity that is different from all other generals as well as distinctly relative to its particular instances. And what is unique may be an actual experience or activity or consciousness, each of which is different from every other experience or act of consciousness.

The minimal condition of newness has been given priority, either explicitly or implicitly, by some writers who claim that the world is everywhere open to creative acts. Some philosophers — such as William James, or those who adopt one aspect of a Whiteheadian view — suggest that every activity or occasion is unique. Each experience, event, and object is new in the sense of being a unique actualized integration of components. As such, it is new in being different. And all things to this extent exemplify creativity. Creativity, then, pervades all things.

It seems to me that universalizing creativity in this way makes the meaning of "creativity" too broad. The price paid for this democratic application is that we have no way to distinguish among creations. With respect to uniqueness in the sense of being different, we do not have even degrees of creativity. A stone exemplifies creativity as much as a painting by Giotto. Except perhaps as a metaphysical category, this universal meaning of creativity does not advance our understanding. It does not help us understand what the creation of a more special or striking kind is. The point that some condition other than uniqueness in the sense of difference is needed can be seen if we notice that "uniqueness" is a term that usually includes more than sheer difference. There must be some merit in being different, and the attribution of the term "unique" implies that it is good to be different. However, the value of being different still is insufficient to justify regarding a valuable and different outcome of an act as a creation in any other sense than the universal, thin sense. For one thing, many of the acts and outcomes which, like all acts and outcomes which in this view are unique, are different in certain *ways* that make them unacceptable candidates as creative or as creations. If this were not so, then the distinction between creative acts and imitation, routine processes, or hackwork would be meaningless.

Now, the *way* singulars are unique, which is to say, the way they are different and therefore valuable, is the key to a second and more radical sense of newness. This point can be considered in terms of the way in which what is unique may be regarded as intelligible.

The most widely accepted view of intelligibility requires that what is intelligible be related to what is familiar. On this view, even if an outcome of a process is different as a singular, its difference is irrelevant to its intelligibility, since it is classified or identified with reference to other things. Thus insofar as it is knowable, the outcome is reducible. In addition, it may also be reducible if it is treated as an instance of regularities and is related to antecedent conditions so that predictions of it are possible. On either alternative, as classified or as an instance of regularity and as predictable, its difference is denied, overlooked, or at least construed as trivial for the purposes of knowledge. Its being different does not contribute to its intelligibility.

Now, according to those who admire uniqueness, the difference noticed in an outcome may be considered thus reducible. Yet, the differences are not denied, overlooked, or viewed as trivial. The outcome is regarded as an individual rather than a repeatable instance. Thus, even though the characteristics of the outcome may be found elsewhere, the outcome is nevertheless significantly different, for the identity of the outcome is said to resist exhaustive characterization. Something is left over. Of course, there are many variations of the interpretation of things as individuals. What is important here, however, is the question whether, on any such interpretation, the differences that mark the individual are peculiar to it so that these differences include or exclude intelligibility. Uniqueness, though not trivial, is regarded as unintelligible because, with respect to uniqueness, the thing is neither describable and sharable through discourse nor predictable, except to the extent of predicting that some individual exhibiting common traits and exemplifying regularities will occur if certain conditions prevail. With respect to what is intelligible, one example will do as well as another. And insofar as the unique thing is intelligible, its newness vanishes.

The minimal condition of newness — irreducible and unintelligible difference — however, is not the only order of newness. Nor is intelligibility exclusively dependent on familiarity. There is a radical and intelligible newness that appears in some outcomes. And it is this order of intelligible newness that is overlooked or left unexamined in so much of the literature on creativity. Failure to give this order of intelligible newness its due has led to too many views that either reduce creativity to natural events or treat creativity as wholly unintelligible and mysterious.

Sometimes there are features of an outcome that do not contribute to its intelligibility as a classifiable and predictable thing. Yet these features contribute to and comprise a complex that constitutes an irreducible intelligible identity. What is irreducible is not an unintelligible element of the outcome. Nor is what is irreducible the bare particularity or discreteness of an individual. What is irreducible is an identity without precedent, an identity not forseeable in terms of repeatable data. The outcome is new in the sense of being different as an intelligibility. This kind of newness is what I have called "Novelty Proper." This is to say that the outcome is novel or new in the proper sense of newness, not in the sense of newness in its lowest common denominator, as a condition universally present in things, but in the sense that is expected when an outcome is regarded as a creation in the radical sense.

New intelligibility is intelligibility that occurs with an identity which appears as a sort of a "Gestalt," that is not reducible to its elements or conditions. The thing regarded as intelligible is an integration of elements, an ordered cluster or matrix whose identity is discernible in a complex of coherent, mutually relevant features. Such an identity appears in what may be called a "structure." The structure is distinctive, individual, and unique. And it is unfamiliar in that it is different from all prior structures. It is irreducible. Yet it is intelligible.

This suggestion, that order or coherence is necessary for unprecedented intelligibility, might be challenged on the ground that creations appear to be incoherent just to the extent that they violate established standards of coherence. Furthermore, many contemporary works of art which are creations seem to depend on incoherence for their import, for they explicitly violate an order that tends toward unity. They thrive on disunity or disintegration.

With respect to the first point, I must agree that creations are in crucial respects incoherent. But they are so in relation to their external contexts. They are in conflict with the standards of coherence in their pasts. However, what is at issue is a coherence that is internal, an order given within the inner context of the creation, within the boundary and focus that distinguish it from other things in the world. Admittedly, if this is new, it must contrast with what was coherent before it appeared.

With respect to the second part of the objection, it should be said that overlooking the order in the internal constitution of created outcomes and emphasizing the relation to their external contexts also lead to the claim that many creations are explicit expressions of disunity. What is taken to be disunity and incoherence in works of art may well be dependent on the unfamiliarity of the creation. To be sure, the order of any outcome that is a creation is elusive. So was the work of Giotto or of Beethoven at one time. But these and all outcomes that call forth response and assessment of them as creations must be based on a recognition of some integrating interaction of their components. There must be at least some inner connections and recurrent components, even if they are only subtly manifest. Without this minimal order, there would be nothing to discriminate. There would be nothing sufficiently determinate to focus attention. Such discriminable focus is what I mean by "coherence."

A new intelligibility, then, is not an instance or example of a general or a universal. Rather, it is a type that is recognizable nowhere else, under no other conditions than those relevant to itself. Its individuality is one of contrast rather than sheer difference of singularity. It contrasts in intelligibility, or in its structural constitution, with all else that was intelligible before it came into being. Thus, the outcome constitutes its own context which contrasts with its outer context of the past. At the same time, the external context is essential to the recognition of newness, for if the outcome were isolated and not seen within a larger context, it would just be itself and not regarded as different, much less a creation.

I hope that my point can be illustrated briefly. Beethoven's Ninth Symphony exhibits a structure that contrasts with his own previously created symphonies. The introduction of the human voice into this orchestral form marks a break with the past. But this break would have been connected with an outcome that is unintelligible if the voice parts were not interwoven with orchestral parts, all contributing to a new focus that makes the qualities of the composition mutually relevant. A new biological species that evolves in nature exhibits a structure that contrasts with structures in its past. The features that sustain it interact with themselves as well as established features so that a different type is established. Over a period of time the new type passes on its structure to off-spring and a new kind is established and functions in a continuing evolutionary process. But the initial occurrence of what becomes a species, even if this occurrence appears gradually, is an occurrence manifest in an outcome that exhibits a contrasting structure.

II

The reason for examining the term "new" in the initial characteri-zation of creativity was to draw out and elaborate criteria of creativity. The first criterion of creativity thus far suggested is that an instance of creativity is found in an outcome that must exhibit a structure which is different from all precedent struc-tures. This formulation of the main criterion has three conse-quences, consideration of which will indicate further criteria.

First, with respect to the outcome, we should notice that since its intelligibility is unprecedented, it is underived and is something that was not predicted. But also, its intelligibility is discernible in a structure that *could not have been* predicted. If the structure were predictable in principle, it would be wholly dependent on prior intelligibility for its own intelligibility. And the structure then would have precedence. It would be derived and prefigured, traceable to antecedents and classified as an instance of general and repeatable conceptualizable items. It would have been known, anticipated, and describable, if sufficient knowledge had been available. To interpret creations this way would be to presuppose some form of determinism, and a determinism which excludes Novelty Proper or newness of intelligibility from the intelligible world. A creation in the radical sense, then, must exhibit structure that is both unprecedented and unpredictable.

The second consequence of the requirement that a criterion must exhibit a structure which is not derived from prior structures was suggested earlier in the observation that those who find newness wherever there is uniqueness seem to include value in their conception of what it is to be unique. Uniqueness is considered good, good aesthetically, morally, or in some ontological or metaphysical sense. What has been said about the coherence that must be present in the ordering of components of the creation similarly suggests value: an epistemological value. Coherence contributes to making the creation recognizable and knowable. It contributes to intelligibility, since it is the basis for the definiteness required for recognizability.

However, the value of coherence pervades all intelligible things. Like the value of uniqueness, this value is too broad. Something more than epistemological value is at stake. If it were not, eccentricities would be admitted as creations. For these exemplify what I am calling "Novelty Proper," possessing some coherence as recognizable structure. But as merely eccentric, they do not merit being considered creations. Being different as a different intelligibility is not sufficient. The outcome must be intelligible in some definite *respect* which is valuable. Its intelligibility must constitute the creation as a model, as an exemplary outcome. This value must initially be inherent. The outcome of a creation is valuable for its own sake. It must exhibit an intelligibility that is good in and for itself. Yet if it contributes to its future, it also has instrumental value. In this connection, it should be acknowledged that not all creations seem to contribute to their futures, at least directly. Bach culminated a tradition, for example. His creations were new structures which, though new, perfected their pasts. On the other hand, El Greco's work awaited a later time to serve the tradition of painting. And some of Picasso's experimental innovations seem to have terminated in strands within a tradition which was not advanced by them. Finally, judgments about the presence of value change. When this happens, either outcomes that were not deemed creations come to be seen as creations, or formerly recognized creations are forgotten or denied status as examples of creativity. They no longer are viewed as contributing to traditions. Yet whenever an outcome is regarded as a creation, it must be regarded as manifesting at least inherent value. It must show itself to be intelligible, as exhibiting structure that is intelligible in some respect that is inherently valuable. Only then could it be exemplary and instrumentally valuable.

The third consequence of the first criterion pertains to the activity that leads to a creation — that is, the creative act. What has been said about outcomes or creations suggests that there is an element of serendipity in creative acts. Such activity is not governed by rules that require, or are required by, the end. As Kant says in *The Critique of Judgment,* the rules are given to art through the talent of the genius. If they were not, the creation would already be complete in principle. The design of the end would be prefigured. And a creative act would be comparable to the process by which an acorn grows into an oak. If one assumes some form of determinism, it might be claimed that creating is simply conforming to a pattern and an envisaged form or structure. Mozart is said to have constructed whole musical compositions without working in any external medium. Thus, if we view the symphony as a mental thing, we might say that it is an envisaged end and that writing a score for it is an act of creation that conforms to its antecedently known structure. But if we say that creations are sometimes completed in the mind of the creator in this way, we do not thereby avoid acknowledging serendipity or the presence of spontaneity in the act. We have simply shifted its locus. We still have before us an act, now "internal," "in the mind," in which the rules constituting it were introduced for the first time and in which an unprecedented structure came to be envisaged by the creator.

The creative act, then, is a transforming act. It proceeds in accord with constraints that are given not only by the personality and environment of the creator and by the medium in which he works, but also in the spontaneity of the act itself. The requirements of the developing structure are themselves developed. The creator does not set out with a pre-envisaged target. The creative artist does not know till he has said it what he wants to say or how he will say it. He constitutes his target as he discovers how to aim at it. Thus, the creative act is discontinuous. It includes at one or more of its stages a break with the constraints that prior intelligibility imposed on the creator. Further, it includes a break in continuities connecting intelligibilities, or structures by which things are duplicable and known before creation takes place.

At the same time, the creative act is not wholly uncontrolled. If there is serendipity, it is not sheer accident. The creator must not only exercise critical judgment in deciding what to accept and reject when possibilities occur to him, but he must also form, refine, and integrate these, even though he knows only with a degree of imprecision what the final integration will be. And most important, he must assume responsibility for what he brings into being. Even automatic writers and the most romantic interpreters of their own creative acts must accept responsibility for the outcome that they believe has its source beyond them.

The responsibility indicates once more the role of value in the creation. For if creators bring their outcomes into being responsibly, they are offered as what ought to be. Moreover, the responsibility of forming an intelligibility which represents a repudiation of what was before acceptable is conducive to the anguish frequently reported by artists. It also confirms why writers like Rollo May speak of the need for courage to create.

With the initial statement of the criteria of creativity and these consequences in mind, the criteria can now be formulated: (1) a creation must manifest what I have termed "Novelty Proper," or a structure that is different and underived from past structures and is thus unpredictable; (2) a creation must be inherently and often instrumentally valuable; (3) the act that leads to the creation includes spontaneity as well as directed control.

III

The adequacy of the criteria I have offered is determined not only by the acceptability of the assumptions and observations that suggest them, but also by their application and their resistance to the challenge of counterexamples. It will be helpful to sketch briefly several applications and possible objections.

First, it might be asked how the criteria apply to Egyptian art. Examples of Egyptian art seem to be repetitive, stylized, and unchanging with respect to their forms. Their intelligibilities seem to be common throughout numerous examples. Does not Egyptian art, therefore, fall outside the criteria? I think it does not. When we regard Egyptian art as creative, we admire it as a general accomplishment seen from the perspective of whole periods. And with respect to a whole period, the form of its examples does contrast with earlier periods. Further, when examples are appreciated as individual works, there are superadded variations from one example to another, variations or contrasts within a common stylized way of depicting, but nevertheless contrasts that show the mark of Novelty Proper. Artists who work according to formulas are regarded as creative because they do not blindly conform to these. They vary the formulas in subtle ways for the sake of approximating or perfecting the formula.

A second application of the criteria suggests that they may be too broad. Consider the experience, for example, of a child learning the multiplication tables. Such learning is not ordinarily viewed as creative. Yet the child's knowledge certainly is new for him. His knowledge consists in apprehending identities for the first time. Is not the child's learning process therefore creative? It is true that in accord with the criteria, there is a sense in which the child might be regarded as creative, though in a very restricted way. He is creative insofar as the novelty in question is severely bounded and personal and insofar as there is directed, deliberate effort coupled with discovery. But the child is hardly creative in the way Poincaré or Newton was. And the criteria do account for this difference. For the child's learning does not introduce spontaneity, the origin of constraints and requirements within the activity of learning; nor does his learning issue in an unprecedented structure. The child's experience can be said to be unprecedented only in the sense of universal all-pervasive novelty according to which each experience can be construed in terms of its uniqueness. But this is not the radical newness or the mark of radical creativity which is at issue.

A third troublesome case for which the criteria are appropriate already has been mentioned. It concerns eccentricities or mere deviations, and acts which are undertaken with the aim of being different for the sake of different. Unfortunately, our admiration for creativity has provoked efforts to be different on the part of would-be creators. Such persons become obsessed, at best, with contributing to society or tradition in some field, or at worst, with enhancing their own self-images. They try deliberately to be different and they offer their eccentricities as creations. The outcomes of their activities may manifest Novelty Proper, insofar as they exhibit unprecedented and unpredictable structures — though they do so less often than their producers would like to think. However, eccentricities lack the criterion of value. They fail to serve as exemplary or as possible models for future work, and they fail to culminate a tradition. A producer of one of them might insist that our blindness prevents our seeing his work as valuable. He may be correct. Nevertheless, until the alleged values of the contrasting structures are recognized, his outcomes will not be deemed creations. But I think it is doubtful that they ever will be. It has been said that an artist sets out, not deliberately to make something beautiful, but rather to express himself. It should also be said that a creative artist sets out not with the intention expressed as "I shall be creative," but rather with the intention to articulate and give form to the elements of a medium. He does this more out of compulsion to work than from the desire to conform to a rule that his work be different. He envisages an outcome that will have determinateness, though it is not yet determined. He envisages an outcome that will be valuable, but not an outcome whose specific value is given. He acknowledges that something he will do ought to be, but he knows not how nor in what respect it ought to be.

There are other applications which further test the criteria of creativity. I do not have space or time to explore these here. However, before concluding, I do want to indicate several philosophical problems raised by acceptance of the criteria. These are problems which need to be dealt with in a theory of creativity that affirms the proposed criteria. First, there is an issue raised by the conception of intelligibility which I have adopted. If intelligibility depends upon the presence of identities, we may ask whether the identities that make things intelligible are immanent to those things. Are they identical with the totality of features relevant to the created outcome? It seems that they must be, since new identities have no prior being or status. If they are new, they must appear for the first time within those things which they make intelligible. This is where they are discovered and created. On the other hand, as identities, they must endure; they must have some constancy so that they are not exhaustively bound by an instant of time and a single place. They must be atemporal as well as temporal in their initial occurrences.

There is a second, related problem. The notion of Novelty Proper is paradoxical. Novelty Proper is exemplified in outcomes that are unfamiliar with respect to their intelligibility. Yet, as intelligible, they must appear *as if* they were familiar. Moreover, what is both intelligible and unfamiliar, though manifest as if it were familiar, is not convertible into something already identifiable that makes the new intelligibility predictable. If it were, newness would be denied in favor of deriving the unfamiliar from what is prefigured.

Now, it should be pointed out that the paradox here is not peculiar to the notion of Novelty Proper. The paradox is simply made evident most sharply and dramatically in the case of creative acts which are instances of Novelty Proper. All learning includes recognition of something not before familiar to the learner. One may insist that such learning is possible because what appears to be unfamiliar is learned because it is referred to the familiar, to a fund of experience that was gradually acquired. But this answer does not account for the first instance of learning, or the encounter with what begins the fund of experience. Something must initially be unfamiliar without a ready-made set of familiar experiences to which it can be traced. Furthermore, the initial recognition of the *connection* that holds between unfamiliar experience to be learned and what is already acquired is itself a recognition of a formerly unknown connection that must occur for the first time.

It should be noted, however, that creative experience is also like learning experience, in that it concerns something unfamiliar initially but which can grow familiar. Unless this were so, our vision of the intelligible world could not grow; what is intelligible could never change.

The third problem to be mentioned centers on the consequence that the criteria I have proposed seems to preclude the possibility of explaining creativity. Creative acts cannot be explained in accord with the traditional requirement that explanations show how things to be explained are predictable. If explanation of creative acts is possible, it must be of a different kind that meets different requirements.

The suggestion I have made elsewhere is that a different kind of explanation is appropriate to creativity. I would call this other kind "an account" rather than "explanation," since the term "account" need not bring with it the standard criteria of explanation. Also, the term "account" is looser in meaning and thus more consonant with what it may accomplish in applying to creative acts. The model for this kind of explanation includes the use of metaphorical expression. Metaphors are used by those who try to speak about creations. Indeed, they seem to be required in order to point us to new intelligibilities. However, more important, metaphors are themselves creations. Consequently, examination of their structure is suggestive of the general character of the structures exhibited in all creations. And if this is so, then metaphors will conform to the criteria of the phenomena for which they are intended to account. Since the purpose of this paper has been to propose criteria of creativity, I must resist embarking on the further task of exploring metaphorical expression.

In conclusion, then, let me reformulate these criteria. Creativity is expected wherever acts in which there is controlled serendipity issue in valuable Novelty Proper. Consequently, the criteria of creativity are: (1) created outcomes have intelligible structures that are irreducible; (2) the structures of created outcomes are unpredictable; (3) the structures of created outcomes are inherently and usually instrumentally valuable; (4) and the acts that lead to created outcomes include an element of spontaneity so that although they are directed and are controlled, they are discontinuous.

Pennsylvania State University

Notes

*The following discussion is a version of a paper presented at a session of The Society for the Philosophy of Creativity, Pacific Division, in conjunction with the American Philosophical Association meetings, Berkeley, California, March 26, 1976. Reprinted from *Philosophy and Phenomenological Research* 40 (1979): 237–49. © University of Buffalo, 1979.
1. Rollo May, *The Courage to Create* (New York: Norton, 1975).
2. Ibid., p. 39.
3. I have tried to take a step in this direction in *A Discourse on Novelty and Creation* (The Hague: Martinus Nijhoff, 1975). This book also develops at greater length, but with different emphases, the criteria suggested in this paper.

THE CREATIVE IMAGINATION*

MICHAEL POLANYI

The enterprise that I am undertaking here has been severely discouraged by contemporary philosophers. They do not deny that the imagination can produce new ideas which help the pursuit of science or that our personal hunches and intuitions are often to the point. But since our imagination can roam unhindered by argument and our intuitions cannot be accounted for, neither imagination nor intuition are deemed rational ways of making discoveries. They are excluded from the logic of scientific discovery, which can deal then only with the verification or refutation of ideas after they have turned up as possible contributions to science.

However, the distinction between the production and testing of scientific ideas is not really so sharp. No scientific discovery can be strictly verified, or even proved to be probable, yet we bet our lives every day on the correctness of scientific generalizations, for example, those underlying medicine and technology. Admittedly, Sir Karl Popper has pointed out that, though not strictly verifiable, scientific generalizations can be strictly refuted. But the application of this principle cannot be strictly prescribed. It is true that a single piece of contradictory evidence refutes a generalization, but experience can present us only with *apparent contradictions,* and there is no strict rule by which to tell whether any apparent contradiction is an *actual contradiction.* The falsification of a scientific statement can therefore no more be strictly established than can its verification. Verification and falsification are *both formally indeterminate* procedures.

There is in fact no sharp division between science in the making and science in the textbook. The vision which guided the scientist to success lives on in his discovery and is shared by those who recognize it. It is reflected in the confidence they place in the reality of that which has been discovered and in the way in which they sense the depth and fruitfulness of a discovery.

D. Dutton and M. Krausz, eds., The Concept of Creativity in Science and Art, pp. 91-108.
© *1981 Martinus Nijhoff Publishers, The Hague/Boston/London. All rights reserved.*

Any student of science will understand—must understand—what I mean by these words. But their teachers in philosophy are likely to raise their eyebrows at such a vague emotional description of scientific discovery. Yet the great controversy over the Copernican system, which first established modern science, turned on just such vague emotional qualities attributed to the system by Copernicus and his followers, which proved in their view that the system was real.

Moreover, after Newton's confirmation of the Copernican system, Copernicus and his followers — Kepler and Galileo — were universally recognized to have been right. For two centuries their steadfastness in defending science against its adversaries was unquestioningly honored. I myself was still brought up on these sentiments. But at that time some eminent writers were already throwing cold water on them. Poincaré wrote that Galileo's insistence that the earth was really circling round the sun was pointless, since all he could legitimately claim was that this view was more convenient. The distinguished physicist, historian, and philosopher, Pierre Duhem, went further and concluded that it was the adversaries of Copernicus and his followers who had recognized the true meaning of science, which the Copernicans had misunderstood. While this extreme form of modern positivism is no longer widely held today, I see no essential alternative to it emerging so far.

Let us look then once more at the facts. Copernicus discovered the solar system by signs which convinced him. But these signs convinced few others. For the Copernican system was far more complicated than that of Ptolemy: it was a veritable jungle of *ad hoc* assumptions. Moreover, the attribution of physical reality to the system met with serious mechanical objections and also involved staggering assumptions about the distance of the fixed stars. Yet Copernicus (*De Revolutionibus*, Preface and Book 1, Chapter 10) claimed that his system had unique harmonies which proved it to be real even though he could describe these harmonies only in a few vague emotional passages. He did not stop to consider how many assumptions he had to make in formulating his system, nor how many difficulties he ignored in doing so. Since his vision showed him an outline of reality, he ignored all its complications and unanswered questions.

Nor did Copernicus remain without followers in his own century. In spite of its vagueness and its extravagances, his vision was shared by great scientists like Kepler and Galileo. Admittedly, their discoveries bore out the reality of the Copernican system, but they could make these discoveries only because they already believed in the reality of that system.

We can see here what is meant by attributing reality to a scientific discovery. It is to believe that it refers to no chance configuration of things, but to a persistent connection of certain features, a connection which, being real, will yet manifest itself in numberless ways, inexhaustibly. It is to believe that it is there, existing independently of us, and that for that reason its consequences can never be fully predicted.

Our knowledge of reality has, then, an essentially indeterminate content: it deserves to be called a *vision*. The vast indeterminacy of the Copernican vision showed itself in the fact that discoveries made later, in the light of this vision, would have horrified its author. Copernicus (Book 1, Chapter 4) would have rejected the elliptic planetary paths of Kepler and, likewise, the extension of terrestrial mechanics to the planets by Galileo and Newton. Kepler noted this by saying that Copernicus had never realized the riches which his theory contained.[1]

This vision, the vision of a hidden reality, which guides a scientist in his quest, is a dynamic force. At the end of the quest the vision is becalmed in the contemplation of the reality revealed by a discovery; but the vision is renewed and becomes dynamic again in other scientists and guides them to new discoveries. I shall now try to show how both the dynamic and the static phases of a scientific vision are due to the strength of the imagination guided by intuition. We shall understand then both the grounds on which established scientific knowledge rests and the powers by which scientific discovery is achieved.

I have pursued this problem for many years by considering science as an extension of ordinary perception. When I look at my hand and move it about, it would keep changing its shape, its size, and its color but for my power of seeing the joint meaning of a host of rapidly changing clues, and seeing that this joint meaning remains unchanged. I recognize a real object before me from my joint awareness of the clues which bear upon it.

Many of these clues cannot be sensed in themselves at all. The contraction of my eye muscles, for example, I cannot experience in itself. Yet I am very much aware of the working of these muscles indirectly, in the way they make me see the object at the right distance and as having the right size. Some clues to this we see from the corner of our eyes. An object looks very different when we see it through a blackened tube, which cuts out these marginal clues.

We can recognize here *two kinds of awareness*. We are obviously aware of the object we are looking at, but are aware also—in a much less positive way—of a hundred different clues which we integrate to the sight of the object. When integrating these clues, we are attending fully to the object while we are aware of the clues themselves without attending to them. We are aware of these clues only *as pointing to the object we are looking at.* I shall say that we have a *subsidiary awareness* of the clues in their bearing on the object to which we are *focally attending*.

While an object on which we are focussing our attention is always identifiable, the clues through which we are attending to the object may often be unspecifiable. We may well be uncertain of clues seen from the corner of our eyes, and we cannot experience in themselves at all such subliminal clues, as for example the effort of contracting our eye muscles.

But it is a mistake to identify subsidiary awareness with unconscious or preconscious awareness, or with the Jamesian fringe of awareness. What makes an awareness subsidiary is *the function it fulfills*; it can have any degree of consciousness so long as it functions as a clue to the object of our focal attention. To perceive something as a clue is sufficient by itself, therefore, to make its identification uncertain.

Let me return now to science. If science is a manner of perceiving things in nature, we might find the prototype of scientific discovery in the way we solve a difficult perceptual problem. Take for example the way we learn to find our way about while wearing inverting spectacles. When you put on spectacles that show things upside down, you feel completely lost and remain helpless for days on end. But if you persist in groping around for a week or more, you find your way again and eventually can even drive a car or climb rocks with the spectacles on. This fact, well-known today, was in essence discovered by Stratton 70 years ago. It is usually said to show that after a time the visual image switches round to the way we normally see it. But some more recent observations have shown that this interpretation is false.

It happened, for example, that a person perfectly trained to get around with upside-down spectacles was shown a row of houses from a distance, and he was then asked whether he saw the houses right side up or upside down. The question puzzled the subject and he replied after a moment that he had not thought about the matter before, but that now that he was asked about it he found that he saw the houses upside down.[2]

Such a reply shows that the visual image of the houses has not turned back to normal; it has remained inverted, but the inverted image no longer means to the subject that the houses themselves are upside down. The inverted image has been reconnected to other sensory clues, to touch and sound and weight. These all hang together with the image once more, and hence, though the image remains inverted, the subject can again find his way by it safely. *A new way of seeing things rightly has been established.* And since the meaning of the upside-down image has changed, the term "upside down" has lost its previous meaning, so that now it is confusing to inquire whether something is seen upside down or right side up. The new kind of right seeing can be talked about only in terms of a new vocabulary.

We see how the wearer of inverting spectacles reorganizes scrambled clues into a new coherence. He again sees *objects*, instead of meaningless impressions. He again sees *real things*, which he can pick up and handle, which have weights pulling in the right direction and make sounds that come from the place at which he sees them. He has made sense out of chaos.

In science, I find the closest parallel to this perceptual achievement in the discovery of relativity. Einstein has told the story of how from the age of 16 he was obsessed by the following kind of speculations.[3] Experiments with falling bodies were known to give the same results on board a ship in motion as on solid ground. But what would happen to the light which a lamp would emit on board a moving ship? Supposing the ship moved fast enough, would it overtake the beams of its own light, as a bullet overtakes its own sound by crossing the sonic barrier? Einstein thought that this was inconceivable, and, persisting in this assumption, he eventually succeeded in renewing the conceptions of space and time in a way which would make it inconceivable for the ship to overtake, however slightly, its own light rays. After this, questions about a definite span of time or space became meaningless and confusing — exactly as questions of "above" and "below" became meaningless and confusing to a subject who had adapted his vision to inverting spectacles.

It is no accident that it is the most radical innovation in the history of science that appears most similar to the way we acquire the capacity for seeing inverted images rightly. For only a comprehensive problem like relativity can require that we organize such basic conceptions as we do in learning to see rightly through inverting spectacles. Relativity alone involves conceptual innovations as strange and paradoxical as those we make in righting an inverted vision.

The experimental verifications of relativity have shown that the coherence discerned by Einstein was real. One of these confirmations has a curious history. Einstein had assumed that a light source would never overtake a beam sent out by it, a fact that had already been established before by Michelson and Morley. In his autobiography Einstein says that he made this assumption intuitively from the start. But this account failed to convince his contemporaries, for intuition was not regarded as legitimate ground for knowledge. Textbooks of physics therefore described Einstein's theory as his answer to the experiments of Michelson. When I tried to put the record right by accepting Einstein's claim that he had intuitively recognized the facts already demonstrated by Michelson, I was attacked and ridiculed by Professor Grünbaum who argued that Einstein must have known of Michelson's experiments, since he could not otherwise have based himself on the facts established by these experiments.[4]

However, if science is a generalized form of perception, Einstein's story of his intuition is clear enough. He had started from the principle that it is impossible to observe absolute motion in mechanics, and when he came across the question whether this principle holds also when light is emitted, he felt that it must still hold, but he could not quite tell why he assumed this. However, such unaccountable assumptions are common in the way we perceive things, and this can also affect the way scientists see them. Newton's assumption of absolute rest itself, which Einstein was to refute, owed its convincing power to the way we commonly see things. We see a car traveling along a road and never the road sliding away under the car. We see the road at absolute rest. We generally see things as we do, because this establishes coherence within the context of our experience. So when Einstein extended his vision to the universe and included the case of a light source emitting a beam, he could make sense of what he then faced only by seeing it in such a way that the beam was never overtaken, however slightly, by its source. This is what he meant by saying that he knew intuitively that this was in fact the case.

We understand now also why the grounds on which Copernicus claimed that his sytem was real could be convincing to him, though not convincing to others. We have seen that the intuitive powers that are at work in perception integrate clues which, being subsidiarily known, are largely unspecifiable; we have seen further that the intuition by which Einstein shaped his novel conceptions of time and space was also based on clues which were largely unspecifiable; we may assume then that this was also true for Copernicus in shaping his vision of reality.

And we may say this generally. Science is based on clues that have a bearing on reality. These clues are not fully specifiable; nor is the process of integration which connects them fully definable; and the future manifestations of the reality indicated by this coherence are inexhaustible. These three indeterminacies defeat any attempt at a strict theory of scientific validity and offer space for the powers of the imagination and intuition.

This gives us a general idea of the way scientific knowledge is established at the end of an inquiry; it tells us how we judge that our result is coherent and real. But it does not show us where to start an inquiry, nor how we know, once we have started, which way to turn for a solution. At the beginning of a quest we can know only quite vaguely what we may hope to discover; we may ask, therefore, how we can ever start and go on with an inquiry without knowing what exactly we are looking for.

This question goes back to antiquity. Plato set it in the *Meno*. He said that if we know the solution of a problem, there is no problem — and if we don't know the solution, we do not know what we are looking for and cannot expect to find anything. He concluded that when we do solve problems, we do it by remembering past incarnations. This strange solution of the dilemma may have prevented it from being taken seriously. Yet the problem is ineluctable and can be answered only be recognizing a kind of intuition more dynamic than the one I have described so far.

I have spoken of our powers to perceive a coherence bearing on reality, with its yet hidden future manifestations. But there exists also a more intensely pointed knowledge of hidden coherence: the kind of foreknowledge we call a problem. And we know that the scientist produces problems, has hunches, and, elated by these anticipations, pursues the quest that should fulfill these anticipations. This quest is guided throughout by feelings of a deepening coherence and these feelings have a fair chance of proving right. We may recognize here the powers of a dynamic intuition.

The mechanism of this power can be illuminated by an analogy. Physics speaks of potential energy that is released when a weight slides down a slope. Our search for deeper coherence is likewise guided by a potentiality. We feel the slope toward deeper insight as we feel the direction in which a heavy weight is pulled along a steep incline. It is this dynamic intuition which guides the pursuit of discovery.

This is how I would resolve the paradox of the *Meno:* we can pursue scientific discovery without knowing what we are looking for, because the gradient of deepening coherence tells us where to start and which way to turn, and eventually brings us to the point where we may stop and claim a discovery.

But we must yet acknowledge further powers of intuition, without which inventors and scientists could neither rationally decide to choose a particular problem nor pursue any chosen problem successfully. Think of Stratton devising his clumsy inverting spectacles and then groping about guided by the inverted vision of a single, narrowly restricted eye for days on end. He must have been firmly convinced that he would learn to find his way about within a reasonable time, and also that the result would be worth all the trouble of his strange enterprise — and he proved right. Or think of Einstein, when as a boy he came across the speculative dilemma of a light source pursuing its own ray. He did not brush the matter aside as a mere oddity, as anybody else would have done. His intuition told him that there must exist a principle which would assure the impossibility of observing absolute motion in any circumstances. Through years of sometimes despairing inquiry, he kept up his conviction that the discovery he was seeking was within his ultimate reach and that it would prove worth the torment of its pursuit; and again, Einstein proved right. Kepler too might reasonably have concluded, after some five years of vain efforts, that he was wasting his time, but he persisted and proved right.

The power by which such long-range assessments are made may be called a *strategic intuition*. It is practiced every day on a high level of responsibility in industrial research laboratories. The director of such a laboratory does not usually make inventions, but is responsible for assessing the value of problems suggested to him, be it from outside or from members of his laboratory. For each such problem the director must jointly estimate the chances of its successful pursuit, the value of its possible solution, and also the cost of achieving it. He must compare this combination with the joint assessment of the same characteristics for rival problems. On these grounds he has to decide whether the pursuit of a problem should be undertaken or not, and if undertaken, what grade of priority should be given to it in the use of available resources.

The scientist is faced with similar decisions. The kind of intuition which points out problems to him cannot tell him which problem to choose. He must be able to estimate the gap separating him from discovery, and he must also be able roughly to assess whether the importance of a possible discovery would warrant the investment of the powers and resources needed for its pursuit. Without this kind of strategic intuition, he would waste his opportunities on wild goose chases and soon be out of a job.

The kind of intuition I have recognized here is clearly quite different from the supreme immediate knowledge called intuition by Leibniz or Spinoza or Husserl. It is a skill for guessing with a reasonable chance of guessing right, a skill guided by an innate sensibility to coherence, improved by schooling. The fact that this faculty often fails does not discredit it; a method for guessing 10% above average chance on roulette would be worth millions.

But to know what to look for does not lead us to the power to find it. That power lies in the imagination.

I call all thoughts of things that are not present, or not yet present – or perhaps never to be present – acts of the imagination. When I intend to lift my arm, this intention is an act of my imagination. In this case imagining is not visual but muscular. An athlete keyed up for a high jump is engaged in an intense act of muscular imagination. But even in the effortless lifting of an arm, we can recognize a conscious intention, an act of the imagination, distinct from its muscular execution. For we never decree this muscular performance in itself, since we have no direct control over it. This delicately coordinated feat of muscular contractions can be made to take place only spontaneously, as a sequel to our imaginative act.

This dual structure of deliberate movement was first described by William James 70 years ago. We see now that it corresponds to the two kinds of awareness that we have met in the act of perception. We may say that *we have a focal awareness of lifting our arm, and that this focal act is implemented by the integration of subsidiary muscular particulars.* We may put it exactly as in the case of perception, that we are focally aware of our intended performance and aware of its particulars only subsidiarily, by attending to the performance which they jointly constitute.

A new life, a new intensity, enters into this two-leveled structure the moment our resolve meets with difficulties. The two levels then fall apart, and the imagination sallies forward, seeking to close the gap between them. Take the example of learning to ride a bicycle. The imagination is fixed on this aim, but, our present capabilities being insufficient, its execution falls behind. By straining every nerve to close this gap, we gradually learn to keep our balance on a bicycle.

This effort results in an amazingly sophisticated policy of which we know nothing. Our muscles are set so as to counteract our accidental imbalance at every moment, by turning the bicycle into a curve with a radius proportional to the square of our velocity divided by the angle of our imbalance. Millions of people are cycling all over the world by skillfully applying this formula which they could not remotely understand if they were told about it. This puzzling fact is explained by the two-leveled structure of intentional action. The use of the formula is invented on the subsidiary level in response to the efforts to close the gap between intention and performance; and since the performance has been produced subsidiarily, it can remain focally unknown.

There are many experiments showing how an imaginative intention can evoke covertly, inside our body, the means of its implementation. Spontaneous muscular twitches, imperceptible to the subject, have been singled out by an experimenter and rewarded by a brief pause in an unpleasant noise; and as soon as this was done, the frequency of the twitches — of which the subject knew nothing — multiplied about threefold. Moreover, when the subject's imagination was stimulated by showing him the electrical effect of his twitches on a galvanometer, the frequency of the twitches shot up to about six times their normal rate.[5]

This is the mechanism to which I ascribe the evocation of helpful clues by the scientist's imagination in the pursuit of an inquiry. But we have to remember here that scientific problems are not definite tasks. The scientist knows his aim only in broad terms and must rely on his sense of deepening coherence to guide him to discovery. He must keep his imagination fixed on these growing points and force his way to what lies hidden beyond them. We must see how this is done.

Take once more the example of the way we discover how to see rightly through inverting spectacles. We cannot aim specifically at reconnecting sight, touch, and hearing. Any attempt to overcome spatial inversion by telling ourselves that what we see above is really below may actually hinder our progress, since the meaning of the words we would use is inappropriate. We must go on groping our way by sight and touch, and learn to get about in this way. Only by keeping our imagination fixed *on the global result* we are seeking can we induce the requisite sensory reintegration and the accompanying conceptual innovation.

No quest could have been more indeterminate in its aim than Einstein's inquiry which led to the discovery of relativity. Yet he has told how during all the years of his inquiry, "there was a feeling of direction, of going straight towards something definite. Of course," he said, "it is very hard to express that feeling in words; but it was definitely so, and clearly to be distinguished from later thoughts about the rational form of the solution." We meet here the integration of still largely unspecifiable elements into a gradually narrowing context, the coherence of which has not yet become explicit.

The surmises made by Kepler during six years of toil before hitting on the elliptical path of Mars were often explicit. But Arthur Koestler has shown that Kepler's distinctive guiding idea, to which he owed his success, was the firm conviction that the path of the planet Mars was somehow determined by a kind of mechanical interaction with the sun.[6] This vague vision — foreshadowing Newton's theory — had enough truth in it to make him exclude all epicycles and send his imagination in search of a single formula, covering the whole planetary path both in its speed and in its shape. This is how Kepler hit upon his two laws of elliptical revolution.

We begin to see now how the scientist's vision is formed. The imagination sallies forward, and intuition integrates what the imagination has lit upon. But a fundamental complication comes into sight here. I have acknowledged that the final sanction of discovery lies in the sight of a coherence which our intuition detects and accepts as real; but history suggests that there are no universal standards for assessing such coherence.

Copernicus criticized the Ptolemaic system for its coherence in assuming other than steady circular planetary paths, and fought for the recognition of the heliocentric system as real because of its superior consistency. But his follower, Kepler, abandoned the postulate of circular paths, as causing meaningless complications in the Copernican system, and boasted that by doing so he had cleansed an Augean stable.[7] Kepler based his first two laws on his vision that geometrical coherence is the product of some mechanical interaction,[8] but this conception of reality underwent another radical transformation when Galileo, Descartes, and Newton found ultimate reality in the smallest particles of matter obeying the mathematical laws of mechanics.

I have described at some length elsewhere some of the irreconcilable scientific controversies which have risen when two sides base their arguments on different conceptions of reality. When this happens neither side can accept the evidence brought up by the other, and the schism leads to a violent mutual rejection of the opponent's whole position. The great controversies about hypnosis, about fermentation, about the bacterial origin of disease, and about spontaneous generation are cases in point.[9]

It becomes necessary to ask, therefore, by what standards we can change the very standards of coherence on which our convictions rest. On what grounds can we change our grounds? We are faced with the existentialist dilemma: how values of our own choosing can have authority over us who decreed them.

We must look once more, then, at the mechanism by which imagination and intuition carry out their joint task. We lift our arm and find that our imagination has issued a command which has evoked its implementation. But the moment feasibility is obstructed, a gap opens up between our faculties and the end at which we are aiming, and our imagination fixes on this gap and evokes attempts to reduce it. Such a quest can go on for years; it will be persistent, deliberate, and transitive; yet its whole purpose is directed on ourselves; it attempts to make us produce ideas. We say then that we are *racking our brain* or *ransacking our brain;* that we are *cudgeling* or *cracking* it, or *beating our brain in trying to get it to work.*

And the action induced in us by this ransacking *is felt as something that is happening to us.* We say that we *tumble* to an idea; or that an idea *crosses* our mind; or that it *comes into* our head; or that it *strikes* us or *dawns* on us, or that it just *presents itself* to us. We are actually surprised and exclaim: Aha! when we suddenly do produce an idea. Ideas may indeed come to us unbidden, hours or even days after we have ceased to rack our brains.

Discovery is made therefore in two moves: one deliberate, the other spontaneous, the spontaneous move being evoked in ourselves by the action of our deliberate effort. The deliberate thrust is a focal act of the imagination, while the spontaneous response to it, which brings discovery, belongs to the same class as the spontaneous coordination of muscles responding to our intention to lift our arm, or the spontaneous coordination of visual clues in response to our looking at something. This spontaneous act of discovery deserves to be recognized as *creative intuition.*

But where does this leave the *creative imagination?* It is there; it is not displaced by intuition but imbued with it. When recognizing a problem and engaging in its pursuit, our imagination is guided both by our dynamic and by our strategic intuition; it ransacks our available faculties, guided by creative intuition. The imaginative effort can evoke its own implementation only because it follows intuitive intimations of its own feasibility. Remember, as an analogy, that a lost memory can be brought back only if we have clues to it; we cannot even start racking our brain for a memory that is wholly forgotten. The imagination must attach itself to clues of feasibility supplied to it by the very intuition that it is stimulating; sallies of the imagination that have no such guidance are idle fancies.

The honors of creativity are due then in one part to the imagination, which imposes on intuition a feasible task, and, in the other part, to intuition, which rises to this task and reveals the discovery that the quest was due to bring forth. *Intuition informs the imagination which in its turn, releases the powers of intuition.*

But where, then, does the responsibility for changing our criteria of reality rest? To find that place we must probe still deeper. When the quest has ended, imagination and intuition do not vanish from the scene. Our intuition recognizes our final result to be valid, and our imagination points to the inexhaustible future manifestations of it. We return to the quiescent state of mind from which the inquiry started, but return to it with a new vision of coherence and reality. Herein lies the final acceptance of this vision; any new standards of coherence implied in it have become our own standards; we are committed to them.

But can this be true? In his treatise on *The Concept of Law,* Professor H.L.A. Hart rightly observes that, while it can be reasonable to decide that something will be illegal from tomorrow morning, it is nonsense to decide that something that is immoral today will be morally right from tomorrow. Morality, Hart says, is "immune against deliberate change"[10] and the same clearly holds also for beauty and truth. Our allegiance to such standards implies that they are not of our making. This existentialist dilemma still faces us unresolved.

But I shall deal with it now. The first step is to remember that scientific discoveries are made in search of reality — of a reality that is there, whether we know it or not. The search is of our own making, but reality is not. We send out our imagination deliberately to ransack promising avenues, but the promise of these paths is already there to guide us; we sense it by our spontaneous intuitive powers. We induce the work of intuition but do not control its operations.

And since our intuition works on a subsidiary level, neither the clues which it uses nor the principles by which it integrates them are fully known. It is difficult to tell what were the clues which convinced Copernicus that his system was real. We have seen that his vision was fraught with implications so far beyond his own ken that, had they been shown to him, he would have rejected them. The discovery of relativity is just as full of unreconciled thoughts. Einstein tells in his autobiography that it was the example of the two great fundamental impossibilities underlying thermodynamics that suggested to him the absolute impossibility of observing absolute motion. But today we can see no connection at all between thermodynamics and relativity. Einstein acknowledged his debt to Mach, and it is generally thought, therefore, that he confirmed Mach's thesis that the Newtonian doctrine of absolute rest is meaningless; but what Einstein actually proved was, on the contrary, that Newton's doctrine, far from being meaningless, was false. Again, Einstein's redefinition of simultaneity originated modern operationalism, but he himself sharply opposed the way Mach would replace the conception of atoms by their directly observable manifestations.[11]

The solution of our problem is approaching here. For the latency of the principles entailed in a discovery indicates how we can change our standards and still uphold their authority over us. It suggests that while we cannot decree our standards *explicitly,* in the abstract, we may change them *covertly* in practice. The deliberate aim of scientific inquiry is to solve a problem, but our intuition may respond to our efforts with a solution entailing new standards of coherence, new values. In affirming the solution we tacitly obey these new values and thus recognize their authority over ourselves, over us who tacitly conceived them.

This is indeed how new values are introduced, whether in science, or in the arts, or in human relations. They enter subsidiarily, embodied in creative action. Only after this can they be spelled out and professed in abstract terms, and this makes them appear to have been deliberately chosen, which is absurd. The actual grounds of a value, and its very meaning, will ever lie hidden in the commitment which originally bore witness to that value.

I must not speculate here about the kind of universe which may justify our reliance on our truth-bearing intuitive powers. I shall speak only of their part in our endorsement of scientific truth. A scientist's originality lies in seeing a problem where others see none and finding a way to its pursuit where others lose their bearings. These acts of his mind are strictly personal, attributable to him and only to him. But they derive their power and receive their guidance from an aim that is impersonal. For the scientist's quest presupposes the existence of an external reality. Research is conducted on these terms from the start and goes on then groping for a hidden truth toward which our clues are pointing; and when discovery terminates the pursuit, its validity is sustained by a vision of reality pointing still further beyond it.

Having relied throughout his inquiry on the presence of something real hidden out there, the scientist will necessarily also rely on that external presence for claiming the validity of the result that satisfies his quest. And as he has accepted throughout the discipline which this external pole of his endeavor imposed upon him, he expects that others, similarly equipped, will likewise recognize the authority that guided him. On the grounds of the self-command which bound him to the quest of reality, he must claim that his results are universally valid; such is the universal intent of a scientific discovery.

I speak not of universality, but of universal intent, for the scientist cannot know whether his claims will be accepted; they may be true and yet fail to carry conviction. He may have reason to expect that this is likely to happen. Nor can he regard a possible acceptance of his claims as a guarantee of their truth. To claim universal validity for a statement indicates merely that it *ought* to be accepted by all. The affirmation of scientific truth has an obligatory character which it shares with other valuations, declared universal by our own respect for them.

Both the anticipation of discovery and discovery itself may be a delusion. But it is fertile to seek for explicit impersonal criteria of their validity. The content of any empirical statement is three times indeterminate. It relies on clues which are largely unspecifiable, integrates them by principles which are undefinable, and speaks of a reality which is inexhaustible. Attempts to eliminate these indeterminacies of science merely replace science by a meaningless fiction.

To accept science, in spite of its essential indeterminacies, is an act of our personal judgment. It is to share the kind of commitment on which scientists enter by undertaking an inquiry. You cannot formalize commitment, for you cannot express your commitment noncommittally; to attempt this is to perform the kind of analysis which destroys its subject matter.

We should be glad to recognize that science has come into existence by mental endowments akin to those in which all hopes of excellence are rooted and that science rests ultimately on such intangible powers of our mind. This will help to restore legitimacy to our convictions, which the specious ideals of strict exactitude and detachment have discredited. These false ideals do no harm to physicists, who only pay lip service to them, but they play havoc with other parts of science and with our whole culture, which try to live by them. They will be well lost for truer ideals of science, which will allow us once more to place first things first: the living above the inanimate, man above the animal, and man's duties above man.

Notes

*Reprinted from *Chemical and Engineering News* 44 (1966): 85–93. © 1966 American Chemical Society.
1. See H. Dingle, *The Scientific Adventure* (London: Pitman, 1952), p. 46.
2. See F.W. Snyder and N.H. Pronko, *Vision with Spatial Inversion* (Wichita: University of Wichita Press, 1952). For fuller evidence and its interpretation in the sense given here, see H. Kottenhoff, "Was ist richtiges Sehen mit Umkehrbrillen und in welchem Sinne stellt sich das Sehen um?," *Psychologia Universalis* 5 (1961).
3. See P.A. Schilpp, ed., *Albert Einstein, Philosopher-Scientist* (New York: Tudor, 1951), p. 53.
4. See A. Grünbaum, *Philosophical Problems of Space and Time* (New York: Knopf, 1963), pp. 378–85.
5. See R.F. Hefferline et al, "Escape and Avoidance Conditioning in Human Subjects Without Their Observation of the Response," *Science* 130 (1959): 1338–39; also R.F. Hefferline, "Learning Theory in Clinical Psychology," in *Experimental Foundations of Clinical Psychology*, ed. A.J. Bachrach (New York: Basic Books, 1962).
6. See Arthur Koestler, *The Sleepwalkers* (London: Hutchinson, 1959).
7. Ibid., p. 334.
8. Ibid., p. 316.

108

9. Michael Polanyi, *Personal Knowledge* (Chicago: University of Chicago Press, 1958), pp. 150–60.
10. H.L.A. Hart, *The Concept of Law* (Oxford: The Clarendon Press, 1961).
11. See Schilpp, op. cit., p. 49.

THE RATIONALITY OF CREATIVITY

I.C. JARVIE

My title links two abstract nouns that are usually set over against each other, seen as contrasting, if not in opposition. The view that informs this paper is that what can usefully be said about creativity is very little, and rather trite; and that it is co-extensive with the rational element in creativity. There may or may not be other than rational elements in creativity; confronted with them, my inclination would be for the first time to invoke Wittgenstein: "whereof one cannot speak, thereof one should be silent." The little I think can be said about the rationality of creativity will be confined to section five. The preceding sections will offer a general critique of the literature, bringing out its poverty and also its irrationality.

Much of the growing literature on creativity,[1] it seems to me, bypasses several quite decisive arguments. Properly understood, these arguments vitiate much of the debate in that literature. These arguments are the following: the problem to be solved by studying creativity is not clearly specified; creativity is treated as a psychological rather than a logical issue; that to explain creativity is to explain it away; and, when we create an explanation that explains creativity, it also, paradoxically, must explain itself. To each of these arguments I shall devote a section.

I. What is the Problem?

What problem is it intended that theories of creativity solve? This, to be sure, is a question rather than an argument. But the absence of a satisfactory answer to the question can be turned into an argument against the enterprise. Einstein remarked in several places that formulating the problem clearly constitutes the better part of the solution. Of course, different writers work on different problems. Were I to *set* problems concerning creativity

D. Dutton and M. Krausz, eds., The Concept of Creativity in Science and Art, pp. 109-128.

for others to work on, I might instance the paradox that the originality of an original idea only becomes "visible" against the background of an intellectual tradition, and those in an intellectual tradition can become trapped in it and blind to new ideas. I might ask whether ideas accumulate, or are reached in one jump. My self-imposed task is not, however, to set problems in this way; it is to express dissatisfaction with a body of literature.

J.P. Guilford,[2] in a paper often cited as initiating the modern literature on creativity, posed two questions for further research. (1) How can we discover creative promise in our children and youth? (2) How can we promote the development of creative personalities? Guilford was, at the time, President of the American Psychological Association, and his setting up of the problems for future research deserves close scrutiny. (1) tells us some children may be creatively promising. (2) implies that there are "creative personalities" (= creatively promising people?), whose development can be promoted. Creativity, then, is envisaged by Guilford as some sort of dependent variable, and the idea is to spot it and get to work on those factors which affect it, in order to foster it.

But what is creativity that we should be interested in discovering it and promoting it? Presumably, it is something like the ability to produce new and original ideas, creations, inventions, and the like. But why should we want to discover thinkers, artists, and inventors, and promote their development? Is it because we want to increase the output of papers, works of art, and patents? Academics, art critics, and patent offices already complain loudly that they can scarcely keep up with the flood. And perhaps much of the flood are ideas, works of art, and inventions that are neither new, nor original, but merely dross. Output alone, then, I take it, cannot be the object of promoting creativity. Obviously, it is output of quality that is to be promoted.

This aim is a rather complex one. Should there be an increase in the output of quality; or that plus a decrease in the output of trash; or merely a decrease in the output of trash? If the second or third, then an overall drop in output as such seems likely; if the first, I would suggest it is impossible − you get more quality only by getting more, and that includes more trash. At the moment, the rate of trash production seems to vary directly with the rate of production. Inverting the relationship might be very desirable. Its accomplishment strikes me as very unlikely, as I shall argue below, because part of the price of maximizing output is a great deal of waste.

Another possible answer to why we want to spot creativity is not because we want to increase it in general, but because we want urgently to foster it in particular cases. Think-tanks, brainstorm sessions, game theory, war games, contingency planning, futurology and economic model-builders all yearn to be creative in this sense; so do cancer researchers, unified field theorists, and governments faced with inflation. All these groups act as though a general theory of creativity might help them to achieve solutions to their problems, and might also help to achieve them sooner. These aims are at once absurdly over-ambitious and perfectly straightforward. It is absurd to imagine us ever being able to solve such pressing problems in a relatively straightforward or mechanical way. Our world does not seem to be made like that. If we routinize a type of problem, as we have in mathematics with tables, logarithms and calculators, that only transfers our ambition from what is now "routine calculation," to deeper and harder problems. The common sense side of the ambition to crack the pressing problems of humanity is heuristics and rules of thumb; general advice for analyzing problems, thinking about them, and so on. These have been forthcoming since intellectual endeavor began. An increase in them looks unlikely, on the face of it, to help with urgent problems.

So far, I have given two answers to the question of why we should want to discover and promote creativity. One is to increase quality output, the other is to facilitate the solution to urgent problems. Both answers are rather pragmatic; to do with practical social gains. Perhaps this is because we have implicitly equated creativity with cognition and technology, rather than with art. We rarely hear cries for an increase in the bulk quantity of art (even of quality art), unless it be from artists; artistic output is rarely acknowledged to be socially urgent (although a case can be made). But for miracle drugs, new sources of energy, even Polaroid movies, there is social urgency. As regards artistic creativity, then, neither output nor urgency constitutes a good reason for investigation. We need a third answer to the question "why research into creativity at all," an answer that is not pragmatic, and that takes account of works of art as well as cognition and technology. My suggestion would be: the simple desire to explain the creative achievement, to understand it better, to get at the truth about it. At the moment, creativity is in the *virtus dormativa* phase: "creativity" is a property ascribed to certain artistic and intellectual achievements by certain creators who are said to possess the capacity to "create," as evidenced by the creativity of their work.

This is a parlous state of affairs indeed — provided only that an explanation of creativity is possible (my scepticism on this score will emerge presently). What, exactly, is in need of explanation; what is the problem? What fact or facts clash with what beliefs, theories, or expectations in the matter of creativity? I confess to not being sure. The obvious explanation of such uncertainty is that different writers have different problems, as we have already noted. There is, however, the possibility that the problems are connected, and hence can be set out in an orderly way. My suggestion for the status of most fundamental problem of all, from which the others branch off, is this. Creative achievements are unique events; explanatory progress is made only with repeatable events. Hence there is something inexplicable about creativity. Or, rather, to all new, interesting, novel, ingenious, pleasing, inventive, gifted: and other cognitive, technological: and artistic acts there is attributed a common property, something they all exemplify, something that can thus be studied and explained: this is called "creativity." Thus the inaccessible uniqueness of creation is made to yield to the belief that nothing is mysterious, remote, inexplicable.

This fundamental problem is, I believe, absorbed during our elementary education. We are taught that artistic, cognitive, and technical achievements are unique events, miracles, strokes of luck (or genius) which we should mainly be concerned to welcome and study.[3] This fundamental epistemological pessimism seems to foreclose the problem: creativity just is an inexplicable "gift." Our rationality is signalled to the extent that we accept and make the best of this situation. That same education, however, also indicates an epistemological optimism: the achievements of science and technology suggest that all problems will sooner or later yield to human thought. We seem to have truth and beauty in such abundance, it is inconceivable that we cannot reach the truth about beauty, originality, etc. Although I claim this problem of inexplicable uniqueness to be the fundamental one behind the creativity literature, I do not claim it is soluble. I suspect that many who contribute to the literature mistakenly believe that because they have found a problem, or what looks like one, it must be soluble. Its insolubility, and indeed ultimate incoherence, in my view vitiates the entire literature.

Given, though, for the moment, that this is a soluble problem, let us continue discussing the attempt to explain creativity. Perhaps if we could explain creativity, specify the laws it obeys and the initial conditions productive of it, then the aims of increased output and/or urgent solutions would be facilitated. The problems of output and urgency would then clearly be subordinate to the problem of inexplicable uniqueness. Were that so, might it not have been better organization on my part to begin with explanation as an aim, and then to move on to increased output and urgent solutions as subordinate or by-product aims? The reason I did not proceed in this way is because of the possibility that an explanation of creativity would not solve the practical problems of output and urgency. Our understanding of earthquakes, continental drift, ice ages, red giants and black holes may make us feel better, emotionally and intellectually, but they do not give us any power over the processes in question. Is it not entirely possible that the concatenation of elements required for creativity obey laws such that human intervention is not possible? Maybe creativity is an outcome of trial and error, and obeys only the laws of chance. We can explain the roulette wheel without being about to gamble on it successfully.

For example, one strand in the literature, which undoubtedly can be traced back to Freud, is the idea that creativity is a manifestation of psychological disorder.[4] The most extreme recent case is Storr's argument that Einstein's genius was a by-product of a psychological condition, namely schizophrenia (could the schizophrenia not be a by-product of his genius?). Confirmations of the neuroses of creative people abound: from Berkeley's concern with the bowels to Einstein's eccentricities of dress and manner; from Bobby Fischer's tantrums to the unbelievable squalor of Francis Bacon's studio. But what of counter-examples? What of

Mozart, Rembrandt, Hume? What of the argument that even if all creators have neuroses, not all neurotics are creative, and hence the creative value of what someone does cannot be fully understood in terms of their psychology? What makes the difference between Einstein's psychopathology and that of the schizophrenic who believes he is a piece of soap? The answer has to focus on the content of their ideas, and thus has to have reference to something like the objective existence of intellectual problems and traditions, and the further fact that Einstein's psychopathology resulted in his contributing to those rather than to the private world of insane fantasy. We cannot, as it were, direct neuroses into science and art rather than intellectual rubbish. This argument goes to undermine the idea that an explanation of creativity, even if found, will lead to practical knowledge of how to foster creativity. We might get further with folk tales and rules of thumb.

So much then for doubts about the utility of solving the basic problem of inexplicable uniqueness. Let me now argue that no such explanations of creativity are there to be found. The case I choose is the movies; not just because they are a speciality of mine, but also because their creative processes are less privatized than those of the traditional arts and sciences (I suppose it is a bit analogous to studying research teams rather than individual scientists). Producers of movies, obviously, have sought for many years to understand what makes a movie a success. Success is an aspect of creativity that cannot be gainsaid simply because popularity is one of its desiderata. In the classical arts, the crucial desideratum of success (so seldom mentioned) was to please the patron, the dedicatee, or God, as well as the artist himself.[5] One might say that in movies, to be a creative success is to make something both interesting *and* popular. By this standard, the most creative periods in Hollywood were between 1914 and 1926, and, again, from the mid-thirties to the mid-forties. The first period marks the maturity and development of the silent movie, the second, maturity and development in the sound movie. What can explain such periods of creative success? Certainly no psychological theory of creativity will do, for this would be to reduce a creative era or a creative place to a fortuitous concatenation of individuals. To give so much to fortune seems excessive.

Two factors are widely bruited:[6] experiment and encouragement. While there may be exceptions, these factors do seem crucial, whether we think of Ingmar Bergman, the French New Wave, or the vintage periods of Hollywood. Freedom to (and security despite) experiment and continuing encouragement are entirely characteristic of Hollywood in the two periods mentioned, less as a policy than as an unintended consequence of institutional arrangements. By 1914, silent movies were twenty years old, technically accomplished, highly successful with the public, and hence not trammelled with rules, regulations, and restrictions. No one really knew for sure and in advance what would click at the box office and be something to be proud of as well. Hence many filmmakers were given their heads as soon as they had a success. This was institutional rather than attitudinal. Doubtless many movie moguls were crass and vulgar and concerned solely with catering to public taste. But they had to work through and with men who had a professional and creative pride; who were not and would not think of themselves as hacks. In some cases their self-estimation may have been excessive, but that is not important. What matters is that men of the caliber of D.W. Griffith, Eric von Stroheim, C.B. DeMille, King Vidor, John Ford, Buster Keaton, and other slapstick comedians were challenged and stimulated by a new medium to an unprecedented degree. At first, they simply enjoyed themselves immensely. Early Hollywood seems to have been a place where any crazy idea could get a hearing, and creators worked at a frantic pace.[7] What sorted out the sheep from the goats was the popular touch, whether in a player or a director. Unawareness may have been something of a blessing, for with public acclaim came self-consciousness and, ultimately, pretension. Griffith and Stroheim are examples, as is Orson Welles in the forties. The most spectacular case, however, is Charlie Chaplin. This London street urchin, who became a brilliant vaudeville comedian of the English music hall, went to Los Angeles to try himself in films in 1913. He worked steadily and quickly until 1916. In that year he was negotiating a new contract with a different company, and this necessitated a trip to New York. He cabled his brother which train he was on, and settled down for the five-day journey. Remarking that in those days he was unrecognized when out of make-up and costume, he recalls being in the middle of shaving when the train was swamped with people in Amarillo, Texas.

They were looking for him! Telegraph operators had spread the word that he was coming, and cities laid on formal welcoming ceremonies.[8] Chaplin himself does not say this, but I detect the beginnings of self-consciousness and decline in his movies from that experience, although there are occasional reversions to form. By 1918, Chaplin persuaded D.W. Griffth, Mary Pickford and Douglas Fairbanks to join him in creating their own film company under the title United Artists. Grandeur indeed.

After the introduction of sound, movies were in effect a new medium. Technical teething troubles over, they conquered the depression business slump and entered another golden age artistically and economically. It sounds incredible, but Los Angeles in the thirties seems to have been a cultural El Dorado. Gathered there were Hugh Walpole, Aldous Huxley, Christopher Isherwood, Thomas Mann, Bertolt Brecht, Thornton Wilder, Scott Fitzgerald, William Faulkner, Raymond Chandler, Nathanael West, Ben Hecht, Igor Stravinsky, Sir C. Aubrey Smith, Sir Cedric Hardwicke, Leslie Howard, Laurence Olivier, Orson Welles, John Houseman, and Vivien Leigh — to name only a few whose fame owed nothing to the movies. The combination of the attractions of California living with the intense creativity and excitement of Hollywood almost produced a cultural center to rival New York. In this atmosphere the individuals sometimes flourished and sometimes suffered, but the movies only flourished.[9]

We can see then that there is some pattern to periods of great creativity. We cannot reproduce them at will. The fundamental strength of public demand for movies cannot be recovered, especially since the arrival of television. Hence, there have been long periods in which filmmakers have been offered anything but encouragement and certainly nothing like freedom. Recently a new phenomenon has appeared: desperate experimentation in the face of bankruptcy. Several of the major American movie companies have gone from success to collapse in the last six or seven years. For a time, there were unrivalled opportunities for innovation and experiment. But that has passed, and a pervading sense of insecurity co-exists with the most profitable year since 1946.[10]

In this section I have argued that the first weakness in studies of creativity is poorly articulated aims. When the aims are articulated it seldom looks as though any of them could sustain the kinds of explorations of creativity that have been undertaken — practical or theoretical.

II. A Logical or a Psychological Issue?

Philosophers commonly distinguish between what they call the logic and the psychology of knowledge; between, that is, the rational reconstruction of the propositions and inferences which constitute knowledge, and the subjective mental processes that go on in those producing it. The aim of making the distinction is to free epistemology from psychology, to deny that knowledge has anything to do with belief. In a similar manner, I should like to distinguish between creativity as an objective matter, a property, so to speak, of created works; and creativity as a property of persons or their minds. Much of the research into creativity is in the latter category and, though I am sceptical of it for reasons already and yet to be stated, I do not want to discuss it directly. What I want to stress is the distinction and the consequences of its neglect. Unless there is an objective logic of creativity, a psychology of creativity can be of no interest. The psychology of scientists is interesting mainly because they are doing science, and science means something like exploring the way the world is. Presumably, creativity is interesting because there is something going on in creative people that results in objective results: creations. What are the objective achievements? Here the whole subject seems to become anecdotal: sayings and stories about the great geniuses are trundled out. Two things strike one: first, the *virtus dormativa* effect; second, that the achievements called creative can be characterized without the concept of creativity and without remainder. Originality, novelty, synthesis, insight, and so on suffice. The status of creativity as a property in its own right is highly questionable.

This is the main reason I have postponed saying anything direct about creativity until the end. So far, we have asked the simple question, what are the aims of studying creativity? This was an obvious question with which to begin a discussion of rationality and creativity, since a rational action is necessarily (though not sufficiently) a goal-directed action. But this question is also a second-order question: it is about the study of creativity, not about creativity. Here we begin to capture in a preliminary way some elements of what this paper is advocating. I believe that the rationality of creativity has to do not so much with the (psychological) creative process, as with the rationality of the creator towards his own creative process. In other words, rationality is to be located in the actions the creator takes, before, and especially after, any "act of creation" occurs.

For various obvious reasons which I do not here want to go into,[11] Popperians often say that rational belief is only a special case of rational action; they subordinate or even absorb the act of belief or creation into action. Hence those who look on creativity psychologically, as bisociation, or psychoneurosis, and so on, as though there was a ghost in the machine, a Magic Moment when the Muse Speaks, a flash of blinding light, an urge to cry Eureka, are bypassed. Action before and action after eventually squeeze out what is in between.

Rational action is goal-directed action, where the tools are critically-held theories. Rational thought is action directed to the goal of solving intellectual or artistic problems. The rationality of rational action shows itself in critical open-mindedness, trial and error. The rationality of rational thought consists of trial and error, which is a form of action. It is not the problem being tackled that is rational; no more is it the solution being offered that is rational; but the method of trying out the thought, inspiration, idea, against the problem, against other solutions and, in some interesting cases, against the facts, that is rational. This is the case whether the problem be called cognitive or artistic.[12]

One hears the response that this is all very well, but that the theory of creativity comes to explain the new thought, the very thing that is put on trial. This presupposes the mind, soul, ghost in the machine, that receives or creates the new thought, the inspiration, does the bisociation, or whatnot. My counter-suggestion would be that my account allows us to dispense with the mind reified in this sense, and hence creativity hypostatized in this sense, for the mind becomes the product rather than the producer of rational organization, of goal-directed, trial-and-error action. Einstein said: "my pencil is more intelligent than I."[13]

This view also proposes a rather different approach to the practical questions of increasing output, solving particular urgent problems. It suggests no recipe, but general social conditions of encouragement and freedom. Bisociation and the like are mental acts which — whether or not they lie at the root of creativity — is somehow irrelevant since there is no way to foster or induce them. Whereas an institutional setup that fosters freedom and encouragement — while not easy to construct — has been constructed from time to time, sometimes inadvertently. The other consequence of my view is that fostering creativity may turn out to be a less attractive prospect than it sounds. What I have in mind is this. Neither a favorable social setup nor an array of rules of thumb, are a recipe for creative achievement. So much expenditure may be undertaken without promise or even hope of results. Indeed, even if success is achieved, they predict that its cost will be a huge amount of what will later be designated waste: that is to say, we know beforehand that much of our expenditure will be afterwards declared waste, yet we cannot specify in advance what will be waste, and hence we cannot eliminate it. Thus we are denied the comfort of even thinking we can eliminate waste. (Although in my view we can dispense with the moral overtones on waste by declaring waste part of the costs of production.) If we divert the entire budget of the U.S. Pentagon into research into the causes of cancer, say, we can no more guarantee success (however that is defined) than we can assure the gambler that his number will come up (it will, but he could be dead before it does). No doubt, in time, scientific research on this scale would yield the bureaucracy, jobs, and contracting industry that the military budget does: and there are unattractive as well as attractive aspects to there being so many men and women in white coats. It is unlikely such expenditure and such waste would ever be agreed to because, while the military has a task at which it is very rarely tested, cancer research is tested every time a patient dies.

So much, then, for the question of what problems are being tackled in the literature on creativity. Not only are these problems not often formulated, formulating them inclines one to the conclusion that they cannot be solved: not simply because they are difficult or intractable but, much more seriously, because they reveal muddles that may vitiate the enterprise. Creativity may not give rise to clear problems because on no known theory is it presupposed as a definite property of the physical world or the world of mind. To avoid concluding it is chimerical we shall have to seek it elsewhere.

III. Creativity a World 3 Event

The groundwork for the second main argument, about to be developed, has been extensively laid down in the discussion so far. Creativity, like knowledge or discovery, can be seen to have a logical as well as a psychological aspect. A solution to a problem can be judged creative; so can the mind which produced that solution. However, a solution that was in fact creative can fail to be recognized as such at the time, and its creator can similarly be thought not creative. The problems we are talking about can be cognitive or artistic. J.S. Bach is my favorite example: a composer virtually dismissed in his own time as a competent but rather pedestrian fellow, far outshone by his brilliant sons C.P.E. and J.C.[14] While those sons are by no means forgotten, they can hardly now be mentioned in the same breath as their father, who has serious claim to being one of the (if not the) greatest composers who ever lived.

With cognition the question becomes more involved. For one thing, there is a popular prejudice in favor of positive science being called creative at the expense of negative science. Positive and negative are connected with the simpleminded distinction between, say, a theory (positive) and its criticism (negative). Sometimes, lord help us, a distinction is made between negative and constructive criticism. Hence, Copernicus is creative for inventing heliocentrism, but the many critics of Ptolemy are not even well-known. Along these lines one might as well say Plato was creative and Aristotle was not: but this shows up the absurdity. Aristotle created beautiful arguments against Plato which were enormous creative achievements in themselves. Newton was a creative genius, but so too was Berkeley, who picked holes in Newton's mathematics and metaphysics.

All this discussion is in the logical realm. Solving a problem is creative; devising a criticism of a theory is creative. Creativeness cannot be unpacked in any precise way, and indeed I have hinted

earlier that it may be a dispensable term. What I want to emphasize here is the different point that whether or not an idea or a
criticism is a creative one has nothing to do with psychology, with
whether an act of creation, inspiration, or bisociation has occurred.
Indeed, the extensive discussion of humor by both Koestler and
Freud demonstrates this. So far as this paper is concerned, there
is nothing creative about jokes. I find the expressions, "what a
creative joke," or, "he's a really creative comedian," very odd.
Odder, is the attention given to humor by those interested in
creativity.

If there is rationality in creativity it is of course going to be
located in the public, objective, logical realm, and not in the
psychological one. Hence the importance of this distinction and
the fact that it plays haywire with the literature on creativity.
On this view, "creativity" is a cognate of "simple," "powerful,"
"unifying," "exciting," "trail-blazing," and other sorts of adjec-
tives that might be attached to new theories, new criticisms,
or new works of art. The rationality of creativity then consists of
it being successful action directed at solving an artistic or cognitive
problem. Calling something "a creative failure," might simply
indicate ingenuity. Degrees of creativity are also sometimes
alluded to, but I see no way of explicating these in any discussible
manner.

Perhaps the notion of creativity, then, is that of a socially
defined event in the world of ideas or art, what Popper has called
the world of objective mind, or mind objectified, the third world
or world 3.[15] This rather than a "property of the mind." A new
theory or work of art undoubtedly has its physical or world 1
aspects — marks on paper or daubs on canvas. It also undoubtedly
in some sense had its origins in certain mental acts of its creator,
in world 2. But neither of these is of much importance or interest.
What makes it a theory, a new theory, a work of art, an original
work of art, is the objective relationships it bears as a world 3
object to other world 3 objects, other theories and works of art.[16]
These in turn have gained their significance because of the bearing
they have on those especially important world 3 entities known as
problems, that is to say, former theories, or means of portraying
the world in art that have now become inconsistent or otherwise
unsatisfactory. The standards of what is new or what is a creative

addition to art are similarly a judgment within world 3, constantly subject to world 3 debate. Art and cognition are the mind striving to achieve objective existence in the external world 3, transcending space and time, and they achieve that when something is created (a theory, a solution, a new problem, a new argument, a "match," a discovery, etc.). Put it like this: to seek in the physical world a recipe for, or an explanation of, what physical conditions bring about Einstein's pencil marks is no more absurd than asking us what kinds of mind or mental process brings about Einstein's great thoughts. Just as Einstein's pencil marks have no meaning unless their significance in the third world is understood, so his thoughts have no interest unless there is a third world significance. This is precisely my objection to Storr's view of Einstein as a psychopath. What possible explanatory value does this hold? What counts is not psychopathy, but the world of objective ideas. This is the logical aspect of creation and cognition; creativity in a certain sense only exists as a vector in world 3.

IV. To Explain Away

The third decisive argument undermining the entire enterprise of studying creativity in the manner it is currently studied has been most eloquently expressed by Einstein in his beautiful essay on Newton.[17] Conceding that reason is weak when measured against its never-ending task, Einstein hails Newton as the first to succeed in explaining a wide range of phenomena mathematically, logically, quantitatively, and in harmony with experience. How, he asks, "did this miracle come to birth in his brain?" Then he apologizes for the "illogical" question, "for if by reason we could deal with the problem of 'how,' then there would be no question of a miracle in the proper sense of the word" (p. 220).

In other words, what Einstein is saying is that creativity is interesting precisely because and to the extent that it is uniquely mysterious. Were it not such, it would not be creativity. We can show this by imagining what the reaction would be were we to pinpoint the creative element in many cases of creation. I contend that what would happen then is that we would set out to isolate other, alternative special features of creativity as "the miracle."

The problem here is that this argument shows that creativity is such that were it to be explained it would be explained away. Usually an explanation of something does not explain it away. When it does, something is wrong.

V. Self-Reference

The fourth decisive argument is this. Suppose we had an explanation of creativity. *A fortiori*, that would be an explanation of new explanations. *A fortiori*, it would be an explanation of that new explanation itself that we have just devised (namely of creativity).

Were we to say, (T) "All creation is a product of neurosis," we should also have to say that (T) itself, being a creation, was a product of neurosis — regardless of who first uttered it. But how are we to know that this neurotic manifestation is to be taken seriously, whereas the belief that someone is a piece of soap is not?

VI. Trial and Error

Having, then, looked at four arguments which seem to me not only not met but quite impossible to meet, the grounds for my scepticism towards the literature on creativity have been exposed. Towards the beginning of the paper I said that what could be said about creativity was rather trite and rather scanty. At the risk of now being both trite and scanty, I present a few remarks on the matter additional to what has emerged in the foregoing discussion. Bertrand Russell, W.A. Mozart, and moviemaking are my examples of creativity; Russell in the cognitive area, Mozart in the area of individual artistic creation, movies in the area of collective artistic creation; Russell and Mozart fluent and individual geniuses, movies made by committees.[18]

In his essay, "How I Write,"[19] Russell reports that, when bothered by a problem, he tried the method of writing and rewriting, but he invariably found the first version to be the best. This was a discovery about himself (his mind, but, more importantly, its world 3 achievements). Eventually, he adopted the following method. After concentrating reflection on the problem at hand, possibly writing nothing, he had planted it in his subconsciousness and would turn to other things. At some point the solution would re-emerge into his consciousness, and it only remained for him to write it out. This is a very striking passage in Russell. It gives concrete and helpful hints ("Turn to other things for a while"). It explains the sense of inspiration or revelation (as the rush created by the surfacing of subconscious mental work). It is also followed by a careful disclaimer. Russell points out that not everything he produced in this manner was creative or valuable or true. He says merely that this was his conscious method of working, which he had discovered by trial and error. After the fact he decided whether what he had written was publishable. Even after he had published he might later decide that some of the things he had said were false. He emphasized, however, that his aim was simply to try to do the best he could at the time. The Russell Archives at McMaster University are filled with early drafts of books and articles that never saw the light of day. Hence we see at work in Russell a conscious process of trial and error, a totally rational approach to creativity as ultimately a world 3 matter.[20]

Similar in some ways is the fluency reported of Mozart, whose subconscious seems to have been working on music the whole time. Hence, he could compose in his head, play from memory, and write down the piece later; compose while playing billiards; even compose a prelude while copying down the already thought-out fugue.[21] He left very few notebooks or sketches, and yet is known to have corrected and improved works a great deal, often turning aside to easier, lighter commissions while struggling hard with problems in more intense works.

I have picked on examples like Russell and Mozart because on the surface they look so fluent and "inspired," the opposite of the hesitant, diffident, groping, trial-and-error creator I have in mind. In close-up, however, this impression disappears. Russell and Mozart have to be rational about their own creativity too. Russell reports a harrowing exception to his own method of work when he was writing *Principia Mathematica*. He tried consciously to work hard and creatively. The strain of working on the abstract and difficult problems of that book from 1902—1910 was so severe that "my intellect never quite recovered . . . I have been ever since definitely less capable of dealing with difficult abstractions than I was before."[22]

Artists and intellectuals work as they can, as they find they have been most successful (in the world 3, not public opinion, sense) and exercise much of their critical faculties on the question of what is good enough to release. Painters throw away canvasses, or store them away incomplete, perhaps to be finished one day, perhaps not; writers submit themselves to editors (who may be themselves, their wives, as well as others), the most startling case of which came to light recently with the publication of Ezra Pound's editing of T.S. Eliot's poem "The Waste Land." Some people went so far as to wonder whether publication didn't detract from Eliot's stature as a creative innovator.[23] The short answer to that, I suppose, is that Eliot could have restored any cuts Pound made had it been his judgment that they did not serve to solve the poetic problem with which he was grappling. Furthermore, many artists and intellectuals show their work to friends, wives, children, professional peers, editors, publishers and the like long before it sees the light of day. Academe has a tradition of acknowledging such help. Because that tradition is not general does not at all imply the practice is not.

It is not hard, then, to see the process of trial and error not only at work, but consciously employed, in cognition and the individual arts. A *fortiori* it is not going to be difficult to detect it in collective arts, whether architecture or the cinema. Filmmaking not only employs trial and error, but, to the extent possible, institutionalizes – even bureaucratizes – it. This is not necessarily all to the good, as, obviously, excessive trial is a form of caution. Once Russell knew that earlier works he had written contained mistakes, or did not sell, he might have been justified in publishing nothing. Capable of error and yet wanting to avoid it, the cautious policy is to do nothing.[24] Capable of failure, yet desperately trying to avoid it, cautious movie producers compromise – that is to say, they interpret trial results too soon and too harshly. This permeates the creative process in movies, which is an institutionalized clash between art and commerce, each side with its own problems, its own integrity, its own series of trials and errors to be avoided. Money being the necessary, but by no

126

means sufficient, condition of there ever being any films made at all, commerce holds a powerful position. The subtlety is not that the bureaucratic organization is split between commercial people and creative people, but that commerce and creation coexist in most film people. Hence, a script may be written and rewritten until it has shape and coherence; it then may be rewritten some more to improve its commercial possibilities. The same will apply to casting, to decisions on sets, costumes and locations, to editing the picture, to the manner in which it is publicized. The writer, the director, the actor, the cameraman, the set designer, and the editor are all simultaneously in search of a solution to their creative problems which is also commercially viable. Some trials are creative (does that scene cut well?), some commercial (does the preview audience like it?), many are a mixture of both. For sure, the box office is neither physical nor psychological; I suggest that the creative side similarly is a sort of objective matter.

In film, as in the other arts, but less so in philosophy and still less so in science, there is a tedious avant-garde with a romantic view of creativity as mainly consisting in not interrupting while the Muse speaks through the favored individual. Life is easier that way. Being rational and self-critical is very hard and sometimes impossible. It is nevertheless the key to anything deserving the label creativity.

York University
Toronto

Notes

1. See J.P. Guilford, "Creativity," *American Psychologist* 5 (1950): 444–54; Arthur Koestler, *The Act of Creation* (London: Hutchinson, 1964); P.E. Vernon, *Creativity* (Harmondsworth: Penguin, 1970); J.W. Getzels and P.W. Jackson, *Creativity and Intelligence* (New York: Wiley, 1962); Anthony Storr, *The Dynamics of Creation* (New York: Atheneum, 1972); and the *Journal of Creative Behaviour,* 1967.
2. J.P. Guilford, op. cit.
3. In the same way that chess masters record and study the grand masters and their great games. No one seems to be trying to find out what makes a great game great and thus deprive it of its mystique. Cf. section 3, below.

4. Storr, op. cit., pp. 61ff. Jean Cocteau somewhere compares being creative to being pregnant. Salvador Dali expresses similar ideas in *The Secret Life of Salvador Dali* (New York: Dial Press, 1961), as does the brilliant and tormented British painter Francis Bacon. See David Sylvester, *Interviews With Francis Bacon* (London: Thames and Hudson, 1975).

5. When I was first introduced to Rembrandt's painting known as "The Night Watch," it was remarked that its bold, active, almost chiaroscuro style had shocked those who commissioned it, and Rembrandt's intransigent artistic temperament on this point contributed to his financial downfall. The idea was that crass patrons always failed to understand geniuses, who had to be tough to maintain their integrity. Later, I noticed this story was declared a legend by contemporary historians, who claim that the picture was always hung, that Rembrandt's commissions did not dry up, and who attribute his downfall to expensive living (especially the house). The theory that all creative artists are misunderstood by hoi polloi is a widespread and pernicious one, since it can be used to rebut all criticism, and even self-criticism and doubt.

6. See J. Agassi, "The Function of Intellectual Rubbish," *Research in the Sociology of Knowledge, Science and Art* 2 (1979): 209–27.

7. Charles Chaplin, *My Autobiography* (London: The Bodley Head, 1964; Penguin edition, 1966), p. 154. Chaplin was amazed when Mack Sennett said to him, "We have no scenario. We get an idea, then follow the natural sequence of events." So different from what he describes as the "rigid, non-deviating routine" of the theatre.

8. Chaplin, op. cit., pp. 175–79.

9. The most evocative account of Hollywood in the thirties and forties is by Charles Higham and Joel Greenberg, *Hollywood in the Forties* (London: A. Zwemmer, 1968). The only book I know which attempts seriously to discuss the creative process in movies without reference to individual genius, but rather to collective endeavor, is Lawrence Alloway, *Violent America: The Movies 1946–1964* (New York: Museum of Modern Art, distributed by the New York Graphic Society, Greenwich, Conn., 1971).

10. See Pauline Kael, "On the Future of the Movies," *New Yorker*, August 5, 1974, pp. 43–59.

11. See J. Agassi, "The Role of Corroboration in Popper's Methodology," *Australasian Journal of Philosophy* 39 (1961): 82–91, and J. Agassi and I.C. Jarvie, "The Problem of the Rationality of Magic," *British Journal of Sociology* 18 (1967): 55–74.

12. What I have in mind is Gombrich's notion of art as "making and matching." See E.H. Gombrich, *Art and Illusion* (London: Phaidon Press, 1960).

13. Einstein is quoted to this effect in K.R. Popper, *Objective Knowledge* (Oxford: The Clarendon Press, 1972), p. 225 (note).

14. I.C. Jarvie, "The Objectivity of Criticism of the Arts," *Ratio* 9 (1967): 67–83. Tovey comments that Bach's art was neglected "as old-fashioned and crabbed" by his younger contemporaries. We owe his rediscovery to Mendelssohn and Schumann. (See "Bach, J.S." in *Encyclopedia Britannica*, 11th edition.)

128

15. K.R. Popper, *Objective Knowledge,* chaps. 3 and 4.

16. This may explain my total opposition (not to modern art but) to modernism in art, the philosophy that traditions must be broken with. How someone trained as, or pretending to be, an artist can even think of this is something of a mystery; it is a bit like an English speaker deciding to utter only gibberish.

17. Albert Einstein, "Isaac Newton," in his *Out of My Later Years* (New York: Philosophical Library, 1950), pp. 219–23.

18. See I.C. Jarvie, *Towards a Sociology of the Cinema* (London: Routledge and Kegan Paul, 1970), chaps. 1, 2, and 3.

19. In Bertrand Russell, *Portraits from Memory* (London: George Allen and Unwin, 1956), pp. 194–97. Storr, op. cit., pp. 42–43 attributes a similar idea to Graham Wallas.

20. To show that Russell's cool detachment has nothing to do with his endeavors being logical and cognitive, one could cite the superb American painter Edward Hopper, who not only creates by "painting, scraping off, and repainting," but also is dominated by the objective logic of creation: "I find, in working, always the distracting intrusion of elements not part of my most interested vision, and the inevitable obliteration and replacement of this vision by the work itself as it proceeds. The struggle to prevent this decay is, I think, the common lot of all painters to whom the invention of arbitrary forms has lesser interest. I believe that the great painters, with their intellect as master have attempted to force this unwitting medium of paint and canvas into a record of their emotions. I find any digression from this large aim leads me to boredom." (Quoted in Lloyd Goodrich, *Edward Hoppper* [New York: H.N. Abrams, 1971], p. 161.)

21. See the account in Erich Hertzmann, "Mozart's Creative Process," in Paul Henry Lang, ed., *The Creative World of Mozart* (New York: W.W. Norton, 1963), pp. 17–30.

22. Bertrand Russell, *The Autobiography of Bertrand Russell, 1872–1914* (London: George Allen and Unwin, 1967), pp. 152–53.

23. T.S. Eliot, *The Waste Land,* a facsimile and transcript of the original drafts including the annotations of Ezra Pound, ed. Valerie Eliot (New York: Harcourt, Brace, Jovanovich, 1971).

24. In the movies this becomes interesting in the political dimension. From time to time Russia and China made it so difficult to avoid ideological error (because of unpredictable shifts in the party line) that filmmaking shrank almost to zero.

CREATIVE PRODUCT AND CREATIVE PROCESS IN SCIENCE AND ART*

LARRY BRISKMAN

In the past few decades, creativity has become rather like money: everyone seems to want more of it. And just as we are living in monetarily inflationary times, so too the notion of creativity has undergone a wholesale devaluation. Soon we shall not only be carting away our weekly salaries in wheelbarrows, but the very act of doing so shall come to be called a creative one. Yet however lax popular standards may become, there seems to me to be one aspect of creativity which will remain constant, and that is that creativity is something *valuable*, and that the notion of creativity is permeated with *evaluation*. To adjudge something to be "creative," in other words, is to bestow upon it an honorific title, to claim that it deserves to be highly valued for one reason or another.[1] Hence, without standards and values, creativity ceases to exist, just as morality ceases to exist. But as with morality, how high (or low) we set the standards is partially a matter for our decision. In this essay, then, I shall adopt a fairly restrictive standard, and in consequence limit the notion of creativity to Science and Art. I hope that by doing so I do not prejudge any important issues, except one upon which I insist: namely, that creativity is something we value, and that the notion of creativity is an evaluative one.

I. Creativity: Possibility or Necessity?

For all our valuing of creativity, it appears to be, not least of all to creative scientists and artists themselves, a kind of mystery, a kind of miracle. Thus, Mozart writes of his best musical ideas: "*Whence* and *how* they come I know not; nor can I force them."[2] In a similar vein, Tchaikovsky writes that "the germ of a future composition comes suddenly and unexpectedly";[3] while Helmholtz reports that his ideas often "arrived suddenly, without any effort

D. Dutton and M. Krausz, eds., The Concept of Creativity in Science and Art, pp. 129-155.

on my part, like an inspiration."[4] Equally, Gauss, in referring to an arithmetical theorem which he had for years tried unsuccessfully to prove, writes: "Finally, two days ago, I succeeeded, not on account of my painful efforts, but by the grace of God. Like a sudden flash of lightning, the riddle happened to be solved."[5] Such quotations could, in fact, be multiplied almost indefinitely; so consider finally, and more lightly, Desmond Morris's recent report that a journalist once asked Picasso: What is creativity? Picasso answered, "I don't know, and if I did I wouldn't tell you."[6]

Now I certainly do not want to claim to know more about creativity than does Picasso, but it does seem to me that the mysteriousness and miraculousness of creativity is, in effect, an important datum about it. It is, I think, something from which we can learn, and which we should try to explain. Yet if we do assume that creativity is a mysterious miracle, then it becomes one of the most mysterious of all miracles — for it is (*pace* Hume) a repeatable miracle. How, then, can we make rational sense of this "miracle"? How, in other words, is creativity possible?

It is, I believe, crucial to see that the problem is to explain the possibility of creativity, *not* its necessity. For if we were to actually succeed in explaining the necessity of creativity, or the necessity of specific creative achievements, then in a sense we would have explained *too much*. To see this, consider what would be involved in such an explanation: take some specific creative scientific or artistic achievement C, and assume that we had some general theory of creativity, or of the creative process, T according to which C was necessary. This would mean that given T, and a description of some relevant set of prior circumstances or initial conditions P, we could actually deduce the attainment of C. But this implies that anyone in possession of T, and given the description P, would have *ipso facto* been in a position to himself create C; and would, moreover be able in principle to simulate the actual creative process of the creator of C. Thus a general theory of creativity, or of the creative process, along the lines of T would provide a kind of recipe for being creative; it would, in effect, provide a set of explicit instructions for attaining creative achievements.

This possibility would, I maintain, have a number of untoward consequences. Firstly, T would rob the actual creator of C of any particular, or individual, claim to creativity, for on the assumption that P was publicly available, T would provide a publicly available, quasi-mechanical, means for creating C. Moreover, T would provide a means for reaching innumerable creative achievements (C', C'', etc.) effortlessly, since all we would have to do would be to enumerate, one after another, statements describing so-far unrealized sets of relevant initial conditions (P', P'', etc.) and then deduce what would be their resultant. By hypothesis, these resultants (C', C'', etc.) would *have to be* creative ones, for otherwise T would not be the type of general theory of the *creative* process (as opposed to some other kind of process) which we are here envisaging. In other words, T would make creativity both too easy and too cheap, since we could have creative achievements for the asking, and this would mean that there would no longer be much point in calling such achievements creative ones. Thirdly, T would turn every creative achievement into something to be expected, given the relevant prior conditions. But this means not only that there need be no surprises, subjectively speaking, but also that there would be *no objective novelty*. For the existence of T would entail that the achievement C was, so to speak, already "contained within" the prior conditions P, and this means that relative to P, C was no novelty. Since, I take it, the novelty of C relative to what preceded it is at least a necessary condition of C's creativity, it follows that T, far from making creativity necessary, would actually make it *impossible*.[7] In essence, then, a theory such as T would eliminate all the mystery and miraculousness of creativity, but it would also eliminate creativity itself.

The above arguments, if correct, go some way towards clearing up what I take to be one of the main mysteries concerning creativity: namely, the mystery of why creativity itself seems to be a mystery. That is, the above arguments against the possibility of a general theory of creativity along the lines of T yield a kind of meta-expectation to the effect that creative achievements will seem mysterious and miraculous even to those who are themselves responsible for them, for in the absence of something like T, creative artists and scientists will be *themselves surprised* by their own creative achievements; even they could not have foreseen or expected them. Moreover, if a theory of creativity of type T is impossible, then there can be no complete explanation of, and hence no explicit set of instructions for, the attainment of creative achievements. But this means that the creative thinker will himself be unable to specify, even in a *post hoc* fashion, *precisely* how he reached his achievement. Hence, his own creativeness will remain a mystery to him in two ways: first, he could not have predicted it; second, he cannot explain afterwards *precisely* how he managed it. Small wonder then that creative thinkers appeal, at the crucial points in their description of their own creative processes, to inspirations, Divine Grace, and so on. Mozart, for example, was thus a better philosopher than he knew when he reported that he did not know whence and how his best musical ideas came, and that he could not force them.

The problem then, as I see it, is emphatically *not* to remove all the miraculousness and mystery from creativity, for that would be to remove creativity itself. Rather, it is to try to explain how creativity is possible, without making it necessary. To adapt Jacques Monod's beautiful idea about the biosphere: we want a theory according to which creative scientific and artistic achievements have the *right* to exist, not one according to which they are under an *obligation* to exist.[8] Hence, how is creativity possible? *Not*, why is creativity necessary?

II. Why the Interest?

I turn, in this section, away from the main problem to a more prosaic aspect of creativity: namely, why have so many thinkers been interested in it? Undoubtedly, different thinkers have different motivations, different reasons for their interest. But it is, I think, instructive to investigate a few typical reasons for this interest, as this will help to bring to the surface some false hopes which, I believe, have been raised by research into creativity. It will emerge, not unexpectedly perhaps, that the philosophers have been on the side of the angels; while the social scientists, and especially the psychologists, have been the villains of the piece. Moreover, their very villainy will actually help us to zero in on our initial problem.

On the whole, then, contemporary philosophers have been concerned with creativity as an aspect of the problem of human freedom. Popper, for example, has argued[9] that no scientific prediction (and hence, no scientific explanation) of the growth of scientific knowledge is possible, on the grounds that we cannot come to know today what we shall only come to know tomorrow. It follows that what Popper calls the world of "objective mind" (or of "theories, arguments, and problems-in-themselves") is an essentially *open* world, in that it cannot contain a theory which will predict the appearance in that world of all future theories. But for Popper this world of "objective mind" (his "3rd World") interacts both with the world of mental states (his "2nd World") and with the world of physical states (his "1st World") — the latter interaction being mediated through the "2nd World."[10] It follows that both the mental world, and more importantly, the physical world, are also open; and this both rules out a purely physical determinism and, in turn, kicks open the door to the possibility of human freedom. In other words, for Popper, the creative potentiality of human thought, the possibility of genuinely creative, unpredictable additions to his "3rd World," opens up the door to human freedom.

As a second example, consider the existentialist philosophers. For them, the notion of human freedom is linked to Sartre's idea of man's being a "Being-for-itself," a being whose existence precedes its "essence" and who is thus condemned to creating his own "essence," rather than a "Being-in-itself," a being whose "essence" is already defined for it. In this way, existentialists too connect the possibility of human freedom to the possibility of human creativity — man is free precisely because he both can and must create himself, because his existence precedes his "essence." Thus, Sartre writes: "If existence really does precede essence, there is no explaining things away by reference to a fixed and given human nature. In other words, there is no determinism, man is free, man is freedom."[11]

Now although I would not wish to endorse a radical existentialist view, the point I do wish to stress is that contemporary philosophers have not, on the whole, seen creativity as something which can be controlled, manipulated, engineered, or predicted. Quite the contrary, they have seen in scientific and artistic creativity an essential element of unpredictability, an area of human thought and action in which men are capable of breaking out of their temporarily self-imposed (or externally imposed) "prisons" into roomier and better "prisons." And the man who can break out of his prison demonstrates his relative autonomy (or freedom) from it.[12]

In part, perhaps, these rather "heroic" views of creativity may stem from the fact that philosophical interest in it has not been pragmatically oriented. The psychologist, however, has enjoyed no such luxury, for psychological interest in creativity has, since the early 1950's, been unabashedly pragmatic in orientation. J.P. Guilford, a highly influential figure in the psychological creativity "movement," writes that "the most urgent reason [for studying creativity] is that we are in a mortal struggle for the survival of our way of life in the world."[13] N.E. Golovin, writing in 1959, warned that unless the United States either increased the number of scientists and engineers she was producing or else enhanced "the average level of creative capabilities of such scientists and engineers" she was bound to lose out in the struggle with the Soviet Union.[14] While the American psychologist T.A. Razik was even more explicit: "In the presence of the Russian threat," he wrote, "'creativity' could no longer be left to the chance occurrences of the genius; neither could it be left in the realm of the wholly mysterious and the untouchable. Men *had* to be able to do something about it; creativity *had* to be a property in many men; it *had* to be something identifiable; it *had* to be subject to the effects of efforts to gain more of it."[15] In other words, creativity could (at least in part) be controlled, manipulated, engineered, and predicted. In fact, it *had* to be.[16]

III. Product, Process, or Person?

Thus convinced of the potential practical utility, and indeed urgency, of research into creativity, psychometricians such as Guilford and Razik, influenced by what could be called "the cult of personality," set out to identify (through factor analysis) a set of personality traits which might be thought to discriminate between creative and non-creative people.[17] The idea behind this, one supposes, is that there is some unique concatenation of traits which go to make up "the creative personality" or "the creative person." Moreover, one assumes, this "creative person" will initiate or carry out (or perhaps even "undergo") some specifiable psychological process — "the creative process" — in order to reach some creative outcome, "the creative product." This, then, suggests the basis for a kind of psychological research programme,[18] which offers us the hope that, *via* an identification of "the creative person" and of "the creative process," we will be able to both explain and foster creativity and, at least in part, predict both *who* is likely to produce creative products (creative persons, who else?) and *when* they are so likely to produce them (when they are initiating or undergoing a creative process, when else?).

And so it would be — if, that is, it were possible to identify creative people or a creative process *independently* of the creative product. In effect, it is just this possibility which I here wish to deny. Clearly, unless we can identify a person as creative, or some special process as "the creative process," independently of the product which is supposed to be the outcome, we cannot predict the appearance of such a product on the basis of theories of "the creative personality" or "the creative process." My claim, then, is that creative people and creative processes can only be identified *via* our prior identification of their scientific or artistic *products* as themselves creative. That is, the person is a creative one and the process was a creative process only in the light of our prior evaluation of the product itself as a creative product. If this is correct, then it follows that a scientific or artistic product is not creative *because* it was produced by a creative person or a creative process, but rather that both the psychological process involved, and the person involved, are deemed to be creative *because* they succeed in producing a product deemed to be creative. It is the creativity of the product which has, so to speak, logical priority. It also follows that any attempted explanation of the creativity of the product on the basis of the creativity of the person, or of his (or her) psychological characteristics or processes, commits the fallacy of *post hoc, ergo propter hoc.* And this implies that the psychological research programme outlined earlier is an incoherent one.

Now the above account, if accepted, clearly raises a crucial question: for if we cannot identify or explain the creativity of a product by reference to the person who produced it, or the process of which it is the outcome, how *can* we identify and explain this creativity? The short answer, I believe, is that we can only identify and explain the creativity of a scientific or artistic product by reference to *prior* scientific or artistic products. That is, a work of art or a scientific theory does not, as it were, wear its creativity on its sleeve ("essentially," or non-relationally); but neither does it gain its creativity by being related, as outcome, to some specific psychological process. Rather, it possesses it in relation to previous artistic or scientific products; its creativity is to be understood and explained in terms of its relation to these prior products. Clearly, such an explanation can *never* have predictive import, for we shall have to be already acquainted with the product whose creativity we want to explain *before* we can explain why it is a creative product. Thus, *even if* we could predict, on the basis of our knowledge of a person's psychology, that he (or she) will produce a certain product, *even then* we would not have predicted, on that basis alone, that he (or she) will produce a *creative* product. For the creativity of the product resides not in its psychological origins, but in its objective relations to other, previous, products. And this, in effect, provides one possible clue to the answer to our initial question: How is creativity possible? For if the creativity of a product resides in its relations to prior products, then creativity might be possible through the critical interaction of the creator with those prior products themselves. And, as might be expected, that such a critical interaction will actually yield a creative product is indeterminate and cannot be predicted. In other words, creativity becomes possible, but not necessary.[19]

IV. Priority of the Product: An Argument

The sceptic (in this instance, the psychologist) may still be unmoved, for I have yet to give an argument for the thesis that the creativity of the product has logical priority over the creativity of the person and his (or her) psychological characteristics or processes. Nor have I sufficiently argued the view that the creativity of the product cannot reside in its psychological origins, but only in its objective relation to prior products. Although these two theses seem to me to be almost obvious, those who still harbour the hope that creativity *must* be fundamentally a property of people and their psychology, and hence something which we can (at least in part) control, foster, and predict, will rightly demand an argument. I shall, therefore, have to oblige them.

A first, and crucial, argument seems to me to be this: in attempting to deal with creativity "scientifically," psychologists are obliged to assume that creativity has the status of a *fact*, that it exists purely naturalistically, in the way that gravity exists, or the moon exists. It is only against such a background that it makes sense to talk of "the creative personality," or of "creative psychological processes," as objects of scientific investigation. Now even if there do exist, as facts, certain psychological processes and personality traits related to creativity, that these are *creative* psychological processes or traits of the *creative* personality is not itself a fact but an evaluation. That is, although we can study certain personality traits and certain psychological processes as facts, we cannot *identify* these personality traits or psychological processes *as creative ones* independently of our standards and values. The reason for this is simple. As I mentioned at the start of this essay, to adjudge some person, process, or product to be "creative" is to bestow upon it an honorific title, to claim that it deserves to be highly prized, highly *valued*, for one reason or another. The question then is: *what* do we evaluate, *to what* do our standards apply, in the first instance? Do we initially evaluate people, processes, or products as "creative"?

As far as I can see, all we have to go on, initially, is a person's *output* (linguistic, scientific, artistic, etc.). For example, would we know anything of the musical "genius" of a Beethoven, or the scientific "creativity" of an Einstein, if both had been totally paralysed, deaf, dumb mutes unable to externalize that genius in objective scientific or artistic products? Clearly not. And what if Beethoven had written only uninspiring pastiche? Or Einstein had never contributed more to science than a best-selling high school physics textbook? Would they still *be* the creative giants we now consider them to be? Again, the answer, I think, is clearly not. In other words, what we evaluate as "creative," that to which our standards apply in the first instance, are a person's products. But since *creative* psychological processes and traits of the *creative* personality do not exist as facts *per se*, but only as the result of an evaluation, and since our evaluation must, in the first instance, be applied to a person's products, it follows that we cannot identify, independently of such products, creative persons or creative psychological processes. Moreover, psychologists, insofar as they study such processes and personality traits as facts *per se*, independently of all evaluations (i.e., insofar as they study them "scientifically"), simply are *not* studying creativity. In other words, a purely "scientific" study of creativity, and *a fortiori* a purely psychological (or even sociological) study of creativity, is impossible.[20]

It is, I think, important to be clear about the above argument. I do not deny (but nor, for that matter, do I assert) that there may be certain independently identifiable psychological processes or personality traits related to creativity. Nor do I deny that these can be studied as facts, which is to say "scientifically." What I *do* deny is that they exist *qua creative* psychological processes, or *qua* traits of the *creative* personality, simply as facts. Rather, that we are dealing with a *creative* psychological process, or a trait of the *creative* personality, is a question of our evaluation, of our standards. Since we can only apply these standards, in the first instance, to the generated products, it follows that we cannot identify the *creativity* of a psychological process, or of a personality trait, independently of the product and of our evaluation of it. In other words, a) the creativity of the product has logical priority over b) the creativity of the person and his (or her) psychological processes, for we cannot identify b) independently of a). And this means that we equally cannot explain a) by making reference to b). So if we *do* want to explain the creativity of some scientific or artistic product *C*, we shall, it appears, have to make reference not to b), but to *C*'s objective relation to prior scientific or artistic achievements. And it is just these two theses which I set out to argue.

V. Priority of the Product: A Further Argument

It might, perhaps, be thought that the above argument is too abstract or conceptual to carry much conviction. So in this section I shall attempt to offer a more "concrete" indication of the priority of creative products over any purely psychologistic account of the creative process (leaving aside now any consideration of the "creative person" or traits of the "creative personality"). Basically, what I shall attempt to argue here is that we cannot even *describe*, let alone understand or explain, the creative process without reference to the products which are its outcome. In other words, even if we could, *per impossibile*, identify some psychological process as a creative one independently of our identification of its outcome (or product) as creative, even then we could not describe this process in the absence of reference to its products, and so could not conceptualize it as being *purely* psychological.

Consider, then, some creative artist (a painter, say) who wants to express some idea, some emotion, some vision; or who desires to solve some artistic problem (for example, the problem of "painting a dark object in the dark"[21]); or who wants to give the illusion of reality. Consider, equally, a theoretical physicist trying to explain some recalcitrant experimental result; or an experimentalist attempting to devise some method for testing a theoretical prediction. It must, I think, be obvious that it would be highly unlikely for the artist to have the completed painting in his mind, all at once, right from the very start; or for the theoretician to have his complete final explanatory structure before him, in his mind's eye, at the very moment he began work on his problem. Rather, one would expect, the artist must build up his painting gradually, stroke by stroke; while the theoretician must build up his conjectural explanation bit by bit (even though he may have got his explanatory "core idea" in a flash). But if this is the case, then it is highly likely that the very thought processes of the artist or scientist will themselves be affected by the work done so far. In other words, the creator, in his very process of creation, is constantly interacting with his own prior products; and this interaction is one of genuine feedback, for the creator is as much influenced by his own initial creations as these were influenced by him. Thus, for example, the scientist's own subjective thinking is itself as much a product of his objective efforts as these are a product of his thinking. While with respect to art, the American painter Ben Shahn makes the same point forcefully: "Painting," he says, "is both creative and responsive. It is an intimately communicative affair between painter and his painting, a conversation back and forth, the painting telling the painter even as it receives its shape and form."[22]

If all this is correct, then it follows that we shall not be able to describe, let alone understand, the creative process unless we make reference to the "intermediary" products which function, so to speak, as "tools" in the creative process itself. In other words, it once again seems that we cannot think of the creative process as being a *purely* psychological process, for our account of that process shall have to make reference not only to the psychological processes of the creator but also to the ("intermediary") products which are themselves helping to *shape* these psychological processes. To put it bluntly: not only are we unable to identify a psychological process as creative in the absence of an identification (or evaluation) of its resultant product as itself creative, but we are unable to identify *any purely psychological process at all as the creative process.* For any such process which we are likely to call "creative" will involve an ineliminable interaction between the creator's psychological states and his own products, with the latter having a genuine feedback upon, and thus themselves actually helping to create, the former. So once again we can see a kind of priority in the product; for although we can describe the creativity of the product without reference to any psychological process (i.e., we can describe it with reference to prior products), we *cannot* describe the creativity of some psychological process without reference to any products. And this is true for two reasons: first, because the process is a creative one *only if* it issues in a product deemed to be a creative one; and second, because any process worthy of the name "creative" will involve an ineliminable interaction with "intermediary" products which themselves help to create the psychological states of the creator.[23]

VI. Aspects of Creative Products

I have, so far, been arguing for a non-psychologistic approach to the problem of creativity. It is, however, time to put a little more meat onto the anti-psychologistic bones, and to outline what I hope is a plausible solution to our main problem: namely, how is creativity possible? As I have already indicated, we cannot identify a process as creative until we have identified (or evaluated) its outcome or product to be creative. So we may say that the creativity of the product will set the *job specification* for any process which we will deem to be creative. That is, until we have answered the question, "What aspects of artistic and scientific products lead us to evaluate them as creative ones?", we cannot answer the question "What kind of process could possibly result in such products?", and so cannot answer our primary question "How is creativity possible?". For creativity is only possible insofar as it is possible to produce creative products. In this section, then, I shall have to concentrate on the former question, and attempt to outline some aspects of creative products. Only afterwards will we be in a position to tackle our main question head-on.

One of the most striking, and in a way, paradoxical, features of great creative advances is how often they appear to be, with hindsight, almost obvious. Einstein puts the point nicely: "In the light of knowledge attained, the happy achievement seems almost a matter of course, and any intelligent student can grasp it without too much trouble. But the years of anxious searching in the dark, with their intense longing, their alternation of confidence and exhaustion, and the final emergence into the light — only those who have themselves experienced it can understand that."[24]

But if creative achievements often seem obvious, "a matter of course," in retrospect, why are they so difficult to achieve? The answer lies, naturally enough, in Einstein's telling phrase about "the years of anxious searching *in the dark*." That is, great creative advances tend to shed light where none has been shed before; one thus cannot approach one's target clearly, because until one has actually reached it, one cannot say precisely what it is (or even if it exists).[25] Popper makes substantially the same point when he compares the quest for new knowledge with the situation of "a blind man who searches in a dark room for a black hat which is — perhaps — not there."[26] Small wonder, then, that creative advances may be simultaneously difficult to make and yet, once made, appear obvious or a "matter of course." For after the advance, we are bathing in its reflected light, and some of that light may very well reflect back onto that very advance itself.

Why, then, are creative advances in science or art the result of such an "anxious searching in the dark"? What features of the creative scientific, or artistic, product are responsible for this? A

142

first step on the way to an answer can, I think, be taken if we recognize that, as a minimum requirement, a creative product must be a *novel* product. Now the notion of novelty requires a background — what is novel against one background may not be novel relative to another background. In other words, the notion of *absolute* novelty is incoherent; we can only judge the novelty of a product by comparing it to those previous products which constitute the background against which it emerged. Thus, at the very least, a creative product must not be contained within, so to speak, the background of prior products, and this partially accounts for the fact that the creator is "in the dark."

But the novelty of a product is clearly only a necessary condition of its creativity, not a sufficient condition: for the madman who, in Russell's apt phrase, believes himself to be a poached egg may very well be uttering a novel thought, but few of us, I imagine, would want to say that he was producing a *creative* one. So if novelty is not enough, what ingredients need be added in order to get creativity? A second requirement, I maintain, is that the novelty must be put to some good purpose. It must achieve some desired or desirable result. In short, it must be *valuable* novelty. My suggestion is that a scientific or artistic product is valuable insofar as it constitutes or incorporates a *solution* to a *problem*; and the notion of a problem, like the notion of novelty, demands a background; what is problematic against one background may not be problematic against another. But this means that the notion of a solution to a problem must be relative to a background as well. Now a *genuine* problem is always one that cannot be solved simply with the available means, or on the basis of the background against which it *is* a problem. In other words, once we see creative products as solutions to problems, we can understand why they *must* be novel; for insofar as a genuine problem only emerges against a background, and cannot be solved on the basis of that background alone, a solution to such a problem *must* constitute a novel addition to that background. Moreover, it also follows that although the creative product is a novel one, a kind of "internal connection" must exist between this novel product and the background of prior products — namely, the "internal connection" of a solution to the problem it solves.[27]

We have thus far reached the following result: a creative scientific, or artistic, product constitutes or incorporates a novel solution to a problem, inherent in a background of prior products, but not soluble on the basis of these prior products themselves. But this is not yet the whole story, for novel problem-solutions may be related to their background in at least two different ways. First, they may simply go beyond, or extend, the existing background. Second, and more interestingly, they may actually *conflict* with this existing background, or necessitate the *modification* of this background. That is, they may improve upon, and thus supplant, parts of this background. In fact, it seems to me that when we are thinking of really great creative achievements (especially in science, although similar considerations can, I think, be applied to certain aspects of art) we are in the main thinking of this latter case. For example, having inherited the problem of unifying terrestrial and celestial physics from the tradition of Galileo and Kepler, Newton actually solves the problem in such a way as to force the modification of that tradition. For his solution conflicts with, and improves upon, the theories of Galileo and Kepler; it supplants them while at the same time explaining them (as limiting cases). Equally, Einstein inherits the problem of reconciling Newtonian mechanics with Maxwellian electrodynamics, and the theory he puts forward solves the problem in such a way as to supercede and replace Newton's theory. In other words, really outstanding creative achievements have a habit of breaking, in important ways, with the tradition out of which they emerged. They, so to speak, *transcend* this tradition. Moreover, insofar as creative products actually conflict with the tradition out of which they emerge, insofar as they are prohibited by that tradition, creative thinking actually involves the thinking of *forbidden thoughts*.[28] No wonder, then, that the creative process involves not only a searching in the dark but also, as Einstein put it, an "*anxious* searching in the dark."

We are now almost in a position to complete our characterization of the creative product. Only one important aspect remains — that of *evaluation*. For a creative product must not only incorporate a novel problem-solution conflicting with the tradition out of which it emerged, it must also be an acceptable problem-solution. That is, it must be evaluated favorably; it must meet certain standards or certain criteria of acceptability. Such standards will, of course, differ for different endeavors, and those applicable in science will

144

quite reasonably diverge from those applicable to art (although there are, I think, more similarities here than are often imagined). Nevertheless, the point remains that before a novel problem-solution can be given the honorific title "creative," it must be evaluated positively as meeting certain standards. Moreover, these standards will normally themselves be incorporated in the background of prior products, in the tradition, against which the problem-solution has emerged.[29] In other words, from the point of view of the background itself, the creative product surpasses that background in a positively evaluated way, in a way meeting certain stringent requirements or standards already inherent in that background.

To sum up this section: a creative scientific, or artistic, product has, I suggest, the following characteristics. First, relative to the background of prior products, it is a *novel* product. Second, it puts this novelty to a desirable purpose by *solving a problem*, such problems being themselves relative to this background and emerging from it. Third, it does so in such a way as to actually *conflict* with parts of this background, to necessitate its partial modification, and to supplant and improve upon parts of it. Finally, this novel, conflicting, problem-solution must be favorably evaluated; it must meet certain exacting standards which are themselves part of the background it partially supplants. Any product which meets all four of these demands I shall call a "*transcendent product*," for it may be said to transcend the background of prior products against which it emerged. In short, then, the thesis of this section can be simply stated: creative scientific or artistic products are transcendent products: they transcend the tradition out of which they sprang.[30]

VII. How is Creativity Possible?

A creative product, then, is a transcendent product. So the question "How is creativity possible?" can be significantly reformulated: How, after all, is it possible to produce transcendent products? In this section I shall have to canvass, all too briefly, I am afraid, a number of different possible answers to this question, and shall use the results of the previous section as the *desiderata* with which any successful theory or model of the creative process will have to cope. In other words, the transcendent nature of creative products shall constitute the *explicandum,* and an acceptable model of the creative process shall have to explain how it is possible for such products to exist.

As a first possibility, consider what could be called the *mechanistic* approach to the creative process. The mechanist tends to see all natural processes (and hence, for him, the creative process) as occurring according to law; he believes that if some thing (or state of affairs) S exists at time t_1 then there must have existed *prior* things (or states of affairs) at an earlier time t_0 which, together with these laws, are responsible for the appearance of S. The mechanist need not necessarily be a determinist (although he usually is), for some of his laws might be probabilistic only. But what he insists upon is that one be able to trace the existence of S back to earlier conditions out of which it arose according to law. Can such a model account for the appearance of a transcendent product C? The answer, I think, is pretty clearly "no." Firstly, the arguments presented in sections 1, 4, and 5 already militate against the mechanist. Secondly, it is difficult to see how such a model could make sense of the novelty of C, for insofar as it tries to explain the appearance of C on the basis of prior conditions, it is trying to trace what is new back to that which is old. Moreover, what *are* the prior conditions against which a transcendent product appears? Seemingly these will consist, in the main, of the background of prior products and especially the problem, emerging out of them, which the transcendent product comes to solve. But this means that the mechanist must be conceiving of the laws governing the creative process, which transform the prior conditions into C, as a kind of problem-solving algorithm; that is, a kind of function which "maps" the prior conditions (including the problem) into a resultant outcome C (incorporating a solution). But if such a problem-solving algorithm actually exists, then there are in fact no genuinely *open* problems (except, perhaps, that of discovering the algorithm!), and so the very possibility of producing transcendent products disappears.[31] We thus reach a conclusion very much in the spirit of section 1: mechanism, far from making creativity possible, would actually rule it out.

Traditionally, at least since the 18th Century, the only alternative which philosophers have seen to mechanism is *randomness*, or pure chance. Moreover, as every geneticist knows, randomness can be a fertile source of novelty. In fact, it almost has to be, since, by definition, pure randomness can result in virtually *anything*, including novelty. So randomness, as a model of the creative process, at least has this in its favour: it can account for the possibility of novelty. But this is about all. For one of the most striking facts about transcendent products is their *appropriateness*, the "internal connection" which exists between these products and the background against which they emerge. This appropriateness is shown not only in such products constituting solutions to problems, but also in the fact that they meet certain exacting standards, and it would, I think, be a bit *too* miraculous if products exhibiting these characteristics were the outcome of pure chance. Moreover, randomness would, in any case, have to be supplemented by something like recognition criteria, or selection criteria; otherwise the creator would either just go on generating products *ad infinitum*, or else he would stop by pure chance. So randomness is not enough; it cannot explain the appropriateness of transcendent products.

The failure of both mechanism and randomness to account for the possibility of creativity has actually led some philosophers to entertain a teleological model of the creative process: that is, the idea that somehow or other the creator is actually being "pulled" into the future by the transcendent product which he has not yet produced. Apart from the fact that I find this idea mildly incoherent, in that it insists upon "the control by the not-yet-there total situation over the present,"[32] it has other weaknesses as well. For in a sense, like mechanism, the teleological account seems to explain *too much*: for if the artist or scientist is being "pulled" into the future by his final creation, why does he ever *fail*? Why is the production of transcendent products so *difficult*? Why all the "anxious searching in the dark"? Moreover, and it seems to me that this is crucial, if the creator is being "pulled" into the future by his final product, how can he ever fail to recognize it once he has reached it? For he seemingly will no longer *feel* any "pull." But there are some dramatic examples of creators actually hitting upon what ultimately turned out to be their final product, only to reject it. Thus Einstein, in 1913 or 1914, in fact generated and considered the "actual field equations only to discard them for what at the time seemed compelling reasons."[33] If these field equations were themselves actually "pulling" Einstein towards them, his discarding of them is well nigh incomprehensible.

But the teleologist does have a point in his favour: the creator *is* being influenced by his products, only by the products he has managed to generate *so far* and not by some hypothetical product which is not yet there. Moreover, his activity *is* goal-directed, purposive, for he is trying to solve a problem. In addition, the creator will himself normally be aware of those very standards which his solution will have to meet if it is to be an acceptable one, and he can thus employ these standards as part of his solution-recognition apparatus. He can thus accept or reject aspects of what he has done so far on the basis of his knowledge of the problem he is trying to solve and of the standards which his solution must meet. In other words, we can say that the creator is under the "plastic" or "soft" control of his *job specification* — of the problem, and of the standards required of a solution.[34]

We are, I think, finally in a position to outline a plausible answer to our question. Basically, the answer rests upon the Darwinian idea of blind variation and selective retention. As should be obvious, Darwinism manages to transcend the limits of a strict mechanism (by recognizing the fertility of blind variation) while avoiding an out-and-out teleology (since these variations are selected rather than directed). And this seems to be exactly what we are looking for — namely, a process combining "blindness" with "control." In other words, I am suggesting that we may plausibly view the creative process on a Darwinian model, as a case of the blind generation of variants coupled with the selective retention of "successful" variants, *all* under the plastic control of the creative job specification.[35]

To elaborate: we want a process which allows for the production of novelty but which is *not random*, for we want to be able to explain the *appropriateness* of the creative product. Moreover, we want a process in which the final product does not itself *direct* its very production — that is, we want a process which is *blind*. However, we also need a process in which the question of production has, in effect, been controlled (plastically) by those very factors which will enable the product to be a transcendent one — in other words, the process, cannot be *too* blind. The only way to satisfy all of these *desiderata* at once, I suggest, is to see the creator as *a*) critically interacting with prior products, with a tradition, so as to put himself "in touch" with problems and standards for acceptable solutions; *b*) generating *blindly, but not randomly*, a hopefully potential solution or fragments of such a solution — *blindly*, because these are generated without foreknowledge of success; *not randomly*, because the generation is itself already under the plastic control of the relevant problem, the background of prior products, and the relevant standards; *c*) critically interacting with this initial, fragmentary product so as to select or reject it, either in part or in its entirety, such a selection procedure being again under the plastic control of the problem, the background, and the relevant standards; *d*) generating blindly, but not randomly, further hopefully potential solutions or fragments, this time under the plastic control not only of the initial problem, the background, and the relevant standards, but also under the plastic control of *what he has already done*, of his initial effort; *e*) repeat step *c* (i.e., critical selection); *f*) repeat step *d* (i.e., blind, but not random, generation); etc. . . . *until,* hopefully, one has managed to generate or assemble a product which either meets the initial job specification or else meets some improved job specification — this improved job specification being itself the outcome of the above process.

It is essential to recognize that if the above schema is to work, the control which the problem, the relevant standards, and especially the background of prior products (the background "knowledge," so to speak) exercise over the creator *must* be a plastic control, not a "cast-iron" one. For the problem, standards, or background "knowledge" may themselves have to be modified during the very process itself. This becomes particularly clear if we remember that often the way to a solution to a problem is actually blocked by some of the background "knowledge," even though the problem owes its very existence *to* this background "knowledge." In this event, the creator will not be able to solve his problem unless he generates a potential solution which actually conflicts with this background "knowledge"; and this helps to explain why our third requirement on transcendent products sometimes *has to be* fulfilled, if some of the other requirements are also to be fulfilled.

Naturally, however, since the above process is blind — i.e., since there is no foreknowledge of a solution — the creator cannot *know* (in advance) that his way *will be* blocked by a particular aspect of the background. But he might, through constant frustration, come to suspect that it is being so blocked, and thus loosen the control that he will allow this background to have over him; that is, he may be driven into becoming a potential revolutionary! Equally, however, his frustration might lead him to suspect that the very problem itself, which he is trying to solve, is in need of radical reformulation, and in *this* case he will loosen the control which his initial formulation has exercised over his generation and selection of variants. Clearly, if the control of either the background or the problem were "cast-iron," it would be impossible for the creator to loosen their control over his generation and selection of variants. In other words, the *plasticity* of the control is crucial to the possibility of creativity, but so is the *control itself*. This is, perhaps, singularly paradoxical and mysterious; but then again, so is what we are trying to explain with its help.

The above account is, I fear, no more than the sketch of a theory, not a full-blown one. But it does, I believe, indicate how it might be possible to produce transcendent products, for it does help us to understand both the novelty and appropriateness of such products, as well as why they may come to conflict with the tradition out of which they sprang. Moreover, such a theory does *not* explain too much, for it clearly does not guarantee success. Nor does it turn creative achievements into something to be expected, or a "matter of course." Just as evolutionary theory does not make *Homo Sapiens* into a necessary being, but does give him the right to exist, so too the theory of the creative process which I have presented gives transcendent products the right to exist, but no existential necessity. It thus leaves ample room for the mystery and miraculousness of great creative achievements; and for the inspiration, luck, or Divine Grace which great creators appeal to in their descriptions of their own creative processes.

VIII. Conclusion: Transcendence and Self-Transcendence

I have been arguing that creative products are what I have termed transcendent products, that they transcend the tradition out of which they spring. I have also been arguing that neither mechanism, nor pure randomness, nor teleology will ever be able to satisfactorily account for the emergence of such products, and so explain how creativity is possible. Rather, I have suggested, we must seek the explanation in a "Darwinian" process of variation and selection which is itself under the plastic control, not of the yet-to-be realized transcendent product, but of the already realized *idea* of such products. But I cannot conclude without a brief mention of what I see to be the implications of all this for the creator himself; for the person who is, after all, ultimately responsible for the creative achievement he has produced.

It is well to remember that the creative scientist and artist is as much the inheritor of a tradition as he is the transcender of one. In a certain sense, he is personally as much the *product* of a tradition as he is the producer of a product which transcends that tradition. In other words, the creative artist or scientist does not simply produce a transcendent product; in a certain sense, he actually transcends *himself.* He produces something which he could not have willed, and which he could not know he had the ability to produce. As the bearer of a tradition, he has not only gone beyond *it,* he has gone beyond *himself*, he has transcended his Self. One is reminded of the beautiful story about Haydn who, listening for the first time to his *Creation*, broke into tears and said: "I have not written this."[36]

University of Edinburgh

Notes

*   Revised, and greatly expanded, version of a paper first presented at seminars in the Universities of Edinburgh and Leeds. I want to thank the participants in those seminars, and especially Stanley Eveling, Leon Pompa, and Geoffrey Cantor, for their critical comments and suggestions. In addition, I owe special thanks to Michael Krausz for a series of stimulating discussions of issues related to this paper. Reprinted from *Inquiry* 23 (1980): 83–106. © 1980 Universitetsforlaget.

1. A similar point has been made by Vincent Tomas in his article "Creativity in Art," reprinted in V. Tomas, ed., *Creativity in the Arts* (Englewood Cliffs: Prentice-Hall, 1964), pp. 97–109.

2. From a letter, reprinted in P.E. Vernon, ed., *Creativity* (Harmondsworth: Penguin, 1970), p. 55. Italics are in the original.

3. From a letter, reprinted in P.E. Vernon, ed., p. 57.

4. Quoted in R.S. Woodworth, *Experimental Psychology* (New York: Holt, 1938).

5. Quoted in Jacques Hadamard, *The Psychology of Invention in the Mathematical Field* (Princeton: Princeton University Press, 1949), p. 15.

6. In. H.A. Krebs and J.H. Shelly, eds., *The Creative Process in Science and Medicine* (Amsterdam: Elsevier, 1975), p. 31. Morris also reports (on p. 58) that a journalist asked Picasso: "What do you think of chimpanzee painting?" and Picasso bit him!

7. I should point out here that, strictly speaking, C would lack novelty relative not to P alone, but only relative to P together with T. But on the assumption either that what T *describes* existed prior to C, or that T *itself* existed prior to C, if follows that, since C is deducible from P plus T, C is no novelty relative to what preceded it, to what existed prior to it.

8. See J. Monod, *Chance and Necessity* (New York: Knopf, 1971), especially Chapter 2.

9. In his "Indeterminism in Quantum Physics and in Classical Physics," *British Journal for the Philosophy of Science* 1 (1950), and elsewhere.

10. For Popper's "3 World" theory, see his *Objective Knowledge* (Oxford: Oxford University Press, 1972), especially Chapters 3, 4, and 6. I should say here that my approach in this paper has been greatly influenced by Popper's "objectivist" epistemology.

11. From J.P. Sartre, *Existentialism* tr. Bernard Frechtman (New York: The Philosophical Library, 1947).

12. *A propos* of this, Popper writes: "At any moment we are prisoners caught in the framework of our theories; our expectations; our past experiences; our language. But we are prisoners in a Pickwickian sense: if we try, we can break out of our framework at any time" ("Normal Science and its Dangers," in I. Lakatos and A. Musgrave, eds., *Criticism and the Growth of Knowledge* [Cambridge: Cambridge University Press, 1970], p. 56). Popper does, I think, tend to underemphasize the *difficulty* of such "breakouts." In fact, I believe that were such "break-outs" easy to achieve, we would no longer see them as creative.

13. J.P. Guilford, "Traits of Creativity," reprinted in P.E. Vernon, ed., pp. 167–88. The quotation is from page 167.

14. N.E. Golovin, "The Creative Person in Science," in C.W. Taylor and F. Barron, eds., *Scientific Creativity: Its Recognition and Development* (New York: Wiley, 1963), pp. 7–23. (The quotation appears on page 8.) This volume consists of selected papers from *three* conferences on "The Identification of Creative Scientific Talent" held at the University of

Utah in 1954, 1957, and 1959, and supported by the U.S. National Science Foundation. This latter fact, together with the fact that Golovin himself served for a time on the White House staff, indicates both the degree of official U.S. support for the study of creativity and the extent to which American psychologists encouraged the belief in the great practical potentialities of their research.

15. T.A. Razik, "Psychometric Measurement of Creativity," reprinted in P.E. Vernon, ed., pp. 155–66. The quotation , with italics in the original, is from p. 156.

16. What, one wonders, is the reasoning behind such optimism? An instructive analogy can, I believe, be drawn between many 20th Century students of creativity and 17th Century students of methodology. For 17th Century thinkers, like Bacon and Descartes, scientific progress was to be guaranteed by "the man of method" whose relentless application of the correct methodology would ensure continuous progress. But we now live in a post-Einsteinian age (in art, a post-Picasso age) and our idea of scientific (and artistic) progress is a more dramatic, revolutionary one. As a result, I conjecture, many 20th Century thinkers have come to see "the man of creative imagination" as the new guarantor of scientific progress. Hence the idea that creativity *had to be* something we could do something about; for otherwise scientific progress becomes a *contingent accident*, a matter of fortuitous happentance. The lesson, I suspect, has yet to be learnt that there is *no* guarantee of scientific progress, and that the fact that we have any knowledge worthy of the name *at all* is itself a miraculously improbable occurrence.

17. The main methodological tool of such studies were the so-called "open-ended tests of creativity." As Liam Hudson has pointed out in his book *Contrary Imaginations* (Harmondsworth: Penguin, 1967), p. 126: "Open-ended tests are known throughout the United States as 'creativity' tests. Yet . . . there is scarcely a shred of factual support for this." For the severe problems involved in validating such tests (that is, in showing that they are actually testing creativity!) see R.J. Shapiro's contribution in P.E. Vernon, ed., pp. 257–69.

18. The notion of a "research program" is due largely to the writings of Popper, Agassi, and especially Lakatos. For a discussion of psychology's dominant modern metaphysical research program – behaviorism – and some suggested reasons for its current demise, see my "Is a Kuhnian Analysis Applicable to Psychology?, "*Science Studies* 2 (1972): 87–97. For further reflections on behaviorism, see my "Skinnerism and Pseudo-Science," *Philosophy of the Social Sciences* 9 (1979): 81–103.

19. At this point, it will be useful to clear up two possible misunderstandings: First, I am not here assuming that a critical interaction with prior products is the *only* way in which creativity might be possible. My point, rather, is that *if* the creativity of a product resides in its relations to previous products, then the *least mysterious* way in which creativity might be possible is via an interaction of the creator with those prior products.

Secondly, my point of view does *not* entail that whenever we adjudge a certain product X to be creative that we are *ipso facto* forced into judging the producer of X to be creative, or the process by which he created X to be a creative process. (Think of the monkey at his typewriter typing out *Hamlet* and you will see why we do not want to be so forced!) In other words, we may very well insist upon additional conditions which have to be met (additional, that is, to the production of a creative product) *before* we will call the producer, or his means of production, creative. In fact, I shall attempt to spell out some of these additional conditions in Sections 5 and 7. What my point of view does entail, however, is that we *cannot* call a person, or his processes, creative without reference to his products; not that we *must* call him, or his processes, creative given such reference.

20. Incidentally, a similar argument can be used to show that a purely "scientific" psychology or sociology of *knowledge* is equally impossible. For "knowledge," like "creativity," is a normative or evaluative notion which applies, in the first instance, to the outcomes or products *of* research and not to the psychological processes or sociological conditions involved *in* research. Thus, we cannot identify as facts *per se* any sociological conditions of knowledge. In other words, far from replacing epistemology, there cannot even *be* a sociology or psychology of knowledge in the absence of epistemology. For in the absence of epistemology (or "the criteriological science of knowledge," to adopt Collingwood's terminology) there simply *is* no knowledge, and so clearly no psychology or sociology *of it. Pace* Quine, epistemology cannot be naturalized. For some rather disastrous consequences of ignoring the evaluative aspect of knowledge, and of blurring the fact/value distinction, see my "Toulmin's Evolutionary Epistemology," *Philosophical Quarterly* 24 (1974): 60–69.

21. The Japanese painter, Yasuo Kuniyoshi, reports (in B. Ghiselin, ed., *The Creative Process: A Symposium* [Berkeley: University of California Press, 1952], p. 55) that this was in fact an artistic problem which he tried to solve. The idea that the artist confronts objective problems is, of course, well known to any reader of Gombrich's brilliant *Art and Illusion* (Princeton: Princeton University Press, 1960). I should mention that in my view an artist's subjective desire to express some idea, emotion, or vision itself constitutes an *objective* artistic problem – namely, the problem of *how* to express it *given* the means at his disposal. Such a problem may, in fact, only be soluble if the artist invents or discovers new means, not previously at his disposal. We might call this, paraphrasing Lakatos, a "creative problem-shift."

22. Ben Shahn, "The Biography of a Painting," in V. Tomas, ed., p. 32.

23. It might here be objected that the fact that creative people talk of "inspiration," "Divine Grace," "sudden and unexpected ideas," "bisociation," and so on when describing their own creative work invalidates my argument. I do not agree – such sudden flashes of "illumination" are invariably preceded by the production of unsuccessful efforts which

themselves help to shape the very "illumination" in question. Moreover, it might be said that examples such as Coleridge's composing of *Kubla Kahn* refutes my suggestion, in that it seems to have emerged full-blown from Coleridge's head without the benefit of any interaction with "intermediary" products. Again, I am not so sure — at the very least Coleridge had to critically interact with his "finished" product before accepting it *as finished*, and this means that before this conscious acceptance the product can be thought of as an "intermediary" one even though it may be identical to the "finished" one!

24. Quoted in Banesh Hoffmann and Helen Dukas, *Albert Einstein: Creator and Rebel* (New York: Viking, 1972), p. 124.

25. This fact about creativity calls to mind what could be called "Plato's Paradox": namely, that one cannot search for something unless one knows what one is searching for. But plausibly one cannot know what one is searching for until one has actually found it. Hence, one cannot search for something until one has found it. In other words, one cannot search for something! This argument is not simply a sophism, for it indicates that one must either have some *imprecise* idea of what one is looking for, or else some *recognitional criteria* for saying that one has found what one is looking for. The importance of such recognitional criteria for creative activity will be discussed in Section 7.

26. Karl Popper, "Replies to my Critics," in P. Schilpp, ed., *The Philosophy of Karl Popper* (LaSalle: Open Court, 1974), vol. II, p. 1061.

27. Interestingly, the idea that a creative product is a novel problem-solution enables us to explain how even *old* ideas may qualify as creative products. For the required novelty is no longer sheer temporal novelty; rather it is the novelty of a solution relative to a problem. And an old idea, invoked in a new problem-situation, could constitute a novel solution. We might, for example, construe Dalton's creativity in this light, for Dalton recognized in the old *physical* ideas of the atomists the potential solution to a pressing *chemical* problem (that of explaining the law of constant proportions) and thus built his "new system of chemical philosophy" upon an atomistic basis. In other words, Dalton did not create the idea of atoms, but he used that idea *creatively*.

28. I owe this way of seeing the thing to Stanley Eveling.

29. It might be thought that this entails that creative contributions to "methodology" or "criteriology" (i.e., the theory of rational standards) is impossible. This is a mistake: a novel methodological idea which solves an outstanding problem in our current methodological theories may actually constitute an improvement from the point of view of the *older* standards themselves. This, in effect, is how the rational improvement of rational standards is itself possible. Agassi has christened this view "the bootstrap theory of rationality" and suggests that of a series of criteria of rationality each can constitute "an improvement on its predecessor by its predecessor's own lights." See Joseph Agassi, "Criteria for Plausible Arguments," *Mind* 83 (1974): 406–16. For further elabora-

tion of, and suggested modifications to, Agassi's idea of bootstrap rationality, see my "Historicist Relativism and Bootstrap Rationality", *The Monist* 60 (1977): 509–39.

30. It should be clear, I think, that the notion of transcendence here developed is not an absolute one — it admits of *degrees*. For differing products may meet the four demands to a greater or lesser extent; or a product may meet only some of the demands and not others. For example, it might meet demands 1, 2, and 4, yet fail to meet demand 3 (i.e., it fails to necessitate any alterations in the background). Thus, from the point of view of this paper it is perfectly sensible to talk of *degrees of creativity* as well. This seems to me to be a crucially valuable result. In fact, I would say that any theory which made creativity into an "all-or-nothing" thing (as Koestler's "bisociative" theory seems to do) should, for that very reason, be rejected.

31. One can, I think, go even further and *prove* that no such problem-solving algorithm (call it "F") can exist. For F would "map" problems into solutions (i.e., Solution=F(Problem)). But problems are themselves functions of the background of prior products (i.e., Problem=f(Background)). Combining these two we get that Solution=F(f(Background)). But solutions often *contradict* this background. How can a function map an x into something inconsistent with x? Only, I suspect, by being itself inconsistent. So no *consistent* problem-solving algorithm of the type envisaged can exist. And, of course, an inconsistent one would be of no use whatsoever, since out of it we could get anything as a solution to anything!

32. E. Vivas, "Naturalism and Creativity," in V. Tomas, ed., p. 90.

33. Banesh Hoffmann and Helen Dukas, p. 119.

34. The notion of "plastic control" is due to Popper. The reason that the control of problems and standards is only "plastic" or "soft" (as opposed to "cast-iron") is that these can themselves be modified by the very attempts to solve them or satisfy them.

35. The psychologist Donald Campbell has already suggested a "Darwinian" model of the creative process in his paper "Blind Variation and Selective Retention in Creative Thought as in Other Knowledge Processes," *Psychological Review* 67 (1960): 380–400. Where my account differs from his, however, is in seeing this "Darwinian" process as under the plastic control of the creative job specification. In a sense, from my point of view, Campbell's process is just a teeny bit *too* blind.

36. Reported in Popper's *Objective Knowledge*, p. 180. The idea of self-transcendence has been beautifully explored in Gombrich's paper "Art and Self-Transcendence," in A. Tiselius and S. Nilsson, eds., *The Place of Value in a World of Facts* (New York: Wiley Interscience Division, 1970), pp. 125–33.

CREATIVITY AS LEARNING PROCESS

C.A. van PEURSEN

I. The process of creative learning

We use a rather simplified model of various levels of learning processes. This model is a comprehensive one in the sense that it tries to combine various theories about learning behavior, theories that often overaccentuate only one aspect of this behavior. A living organism is always responding to the stimuli that come from the environment and that have a certain influence on such an organism. It is not difficult to differentiate among various relationships, various patterns of stimuli plus the responses resulting from those stimuli. A certain stimulus, for instance the movement of an object towards the eye, produces a fixed type of response, in this case the shutting of the eye. Such a correlation between a certain stimulus and a certain response is called a reflex. All organisms possess an inventory of fixed reflexes. These are inborn, or, in other words, defined by heredity. The stimuli operate as signals that evoke a certain response from the organism. Those patterns of hereditary reflexes protect the organism against dangerous changes in its environment.

A new element in animal behavior comes to the fore in learning processes. There exists no fixed relationship here between a certain stimulus and a response. The organism can produce a response which it did not produce before on similar occasions. New reflexes are built into the pattern of behavior. This means that the range of action becomes enormously enlarged. Beside the reflexes given by heredity, there are now also newly acquired reflexes. The higher an organism is structured, the more complicated its central nervous system, the wider its field of action thanks to higher levels of learned behavior.

A first, lowest, level of learning behavior is found in the conditioned reflexes. This type of learning, also called "classical con-

D. Dutton and M. Krausz, eds., The Concept of Creativity in Science and Art, pp. 157-185.

ditioning," was studied in the well-known experiments of the Russian physiologist I.P. Pavlov (1849–1936). A dog is kept in a laboratory. Food is brought in and the dog starts salivating. At the same time there is a ring of a bell. After some repetition, the dog starts salivating as soon as it hears the bell ringing. In this case, the original stimulus, the smell of the food, has been replaced by a new signal, the ringing of the bell.

Since Pavlov, experiments have gone much further. Dogs have learned to see the difference between mathematical figures like a circle and an ellipse. When the focal distance in the ellipse is gradually reduced there comes a moment that the dog cannot anymore distinguish both figures — the circle and the ellipse — and then he becomes aggressive. A special therapeutic treatment is necessary, consisting among other things of presenting to the dog very easy tasks, in order to make him again suitable for his scientific career as experimental dog; just as with human beings, one would be inclined to say. And indeed, sometimes the whole of human learning-behavior is explained on basis of these dog-like conditioned reflexes with the reinforcements of punishment and remuneration. M. Polanyi refers to these theories and mentions the case of a trained dog who could distinguish the difference between an X-ray photo of healthy dog lungs and an X-ray photo of infected dog lungs. The conclusion was that the dog had learned to distinguish between healthy and sick lungs. Polanyi states that this conclusion is a wrong one. The dog does not understand the X-ray photo, neither does he show an insight in geometry. He has only learned to discriminate between stimuli that are followed by an award and stimuli that are not. Human learning cannot fully be explained on basis of such conditioned reflexes. The stimuli replace original stimuli and the new stimuli have been associated, under pressure of the circumstances (reinforcement), with the original ones. Therefore, we can call the new stimulus a sign of the original one. But this sign has no meaning of its own. It is, in other words, not functioning as a symbol. Insight into lung disease and into geometrical rules is completely absent.

Pavlov thought that this type of learning was related to the central nervous system. It is, but it can also occur, we know now, in animals without such a centralized nervous system. Planaria have also been trained in laboratories to recognize new stimuli and Arthur Koestler has written an interesting account of their operations. This first level of learning behavior is the lowest, as animals without a central nervous system can perform it. It consists of the possibility of accepting a new stimulus as signal or sign of a stimulus that originally was hereditary.

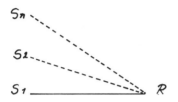

A second level of learning behavior is constituted by operant conditioning. The name of B.F. Skinner is especially connected with research into this type of learning. Here the substitution which takes place is not the substitution of the stimulus but of the response. In other words, here a special activity of the learning animal is required. Well-known examples are rats that try to find their way through a maze in order to obtain food at the exit, or cats that try to reach food by pressing a little lever in order to open the door of the cage. First haphazardly, and then after a long process of trial and error, more and more decisively, those animals learn to solve problems. The experiments can be made more complex by letting the animal choose between little doors with different colors, afterwards between differently shaped doors, or even by mixing both systems of choices and by measuring the time required to relearn new patterns of behavior.

As has been said before, in this case an initiative from the animal is required. Different from a substitution of a stimulus, which must be awaited, the animal by substituting his own response can evoke the desired situation. This type of behavior is only possible in animals that possess a relatively highly organized central nervous system. It constitutes therefore a higher level of learning than that of the conditioned reflex. Some authors, again, try to explain all learning behavior on the basis of these forms of operant conditioning. Skinner himself, for instance, tried to explain human speech on the basis of substitution of responses, a substitution which, via a process of trial and error and guided by reinforcements, acquires in the long run a fixed pattern. This type of reducing all learning behavior to one level has also met with severe opposition. It is well known how N. Chomsky disputed this issue with Skinner.

A third, still higher, level of learning behavior can be distinguished. We call it the semantic conditioned reflex, which means it constitutes linguistic behavior. Pavlov saw this distinction already. In his conditioned reflexes new stimuli, new signals, were learned. But there exist also, said Pavlov, signals of signals: the words of a language. Instead of the mere ringing of a bell one could use the word "ring" and that word would act as the sign indicating another signal. The word does not act in itself as signal; then it would be merely like a cry of warning. Now the word has an influence via its reference, its meaning. The modern investigator A.R. Luria characterizes this system of semantic reflexes as making it possible to bypass a whole series of learning processes. It is clear that by words one can catalogue many objects and events, and even replace various acts. A discussion of how to bridge a river can be very economical and prevent a lot of useless actual behavior. Trial and error through words is a short cut compared with the elaborate actions of animals in a trial and error situation.

Such a system of semantic conditioned reflexes is actually wider than linguistic behavior. It does not include only the use of language but also the use of any system of symbols whatsoever. Many authors such as John Dewey and Ernst Cassirer distinguish sharply between signs and symbols. Signs show a natural link with the things they are token of. Where smoke is you will find fire, smoke being a sign replacing the original stimulus fire. Like the ringing of the bell for the learned dog is a sign substituting the smell of food. But a symbol is, as Dewey formulated, an artificial sign. It does not possess a natural link with the designated object or state of affairs, but has a specific meaning which can change and about which one can enter into conventions and agreements.

Using smoke in order to send a message in Morse code is not a matter of making use of a natural link between smoke and fire. It is a symbolic way of expression which can vary from culture to culture and even from individual to individual. Also artistic expression is more than using only vital signals; it is an effort of communication through symbolic means.

When we try to schematize this type of learning behavior we must conclude that here the stimuli as well as the responses have been made replaceable. By symbols and especially by language one can give various stimuli. Symbols can create a world of anxiety on a quiet sunny afternoon. Symbols can also replace responses. Instead of fighting, communication can, at least in many cases, avoid war as its courteous and well-polished vocabulary often presents a subtle way of making war. This type of learning behavior combines the two lower levels, making both stimuli and responses variable. It is found only in the animal with the most highly organized central nervous system, that is in man. We get in this way a whole network of substituting relations among stimuli and responses. This network represents actually a pattern of human behavior which possesses more freedom, wider ranges of choices. The symbolic network makes it possible to postpone actions, to create distance between a stimulus and a response, to prepare actual behavior by internalized behavior. What we call the process of thought and deliberation might be explained as the growing distance of the symbolic network enlarging human power of control over the surrounding world. Language, and all symbolic behavior, like the ritual dance of warriors, enables a suspension of direct action and creates a preparatory phase of reflection and social communication.

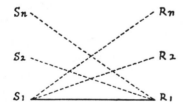

This brings us to the fourth and highest form of learning behavior. A language and a group of symbols constitute a coherent system. Symbols and words get their meaning from a context;

they never operate as separate entities but obey certain rules, certain prescriptions as to their use. The term "network," used before, points already in this direction. This context of rules will be indicated here by the term "grammar," not only in the narrow sense of the grammar of a certain language, but in the wider sense of the rules regulating every coherent system of symbolic behavior.

The symbols that can replace stimuli as well as responses constitute a network that acts as a certain equilibrium, a kind of steady state. This means that they are as much an adaptation to, as an assimilation of, the surrounding world. In this way, human behavior acquires the widest range of possibilities. But this human behavior, like any type of animal behavior, is under constant pressure from the surrounding world. Unexpected events, newly arising situations, threats, and chances are as much challenges to renew the existing pattern of behavior. Traditions must be changed, scientific theories reshaped, religious interpretations have to acquire a new signification, political systems must be restructured. In some cases it might be sufficient to add new symbols to the existing vocabulary or inventory. Sometimes the new symbols are not only added to the old ones but have to replace some of them. But in many newly arising situations this appears to be not sufficient. More profound changes are necessary, perhaps to replace the whole grammar of the symbolic field. And this is actually the fourth level of learning, which changes the whole pattern of behavior as it substitutes not only stimuli and responses, but the grammar itself underlying the network of stimuli and responses.

The substitution of a grammar by a new one does not necessarily imply the substitution of separate symbols. The same symbols can get a new meaning when their frame of reference is being renewed. Many examples could be given. The seemingly simple activity of measuring velocity gets a new signification if performed not within the grammar of Newtonian physics but within that of the special relativity theory of Einstein.

Accentuating the eyelashes by Egyptian women thousands of years ago, when compared with the same operation for a modern woman, acquires quite a new signification. And this because of the change in grammar: in Egyptian time one of status related to a feudal society, in modern times one of a more psychological status related to a highly internalized society where the impersonal symbolic personality is communicated through mass media. In contemporary political speech the same words have a quite different impact than twenty years ago. Sometimes even well-known symbols are explicitly used in their old sense but linked in a rather surprising way in order to evoke the idea of a completely new grammar not known as yet to those who listen to this new associative type of language; as an example I quote a simple question put by pupils of a secondary school to each other: "name a color under ten" — and the answer: "Thursday!"

This last example is not so absurd as one would be inclined to think. It points to the most characteristic feature of this fourth level of learning processes. We shall call this process "restructuring," which means that a new grammar restructures a whole symbolic field. And the feature of this restructuring is inventiveness or even creativity. Creativity is often considered to be an irrational phenomenon. But it is part of human learning, although not in the sense that one has to acquire a piece of new information, but a new frame of reference. Therefore new grammars or structures often make, in the beginning, the impression of absurdity. It is not finding new words or new symbols, but to give new meanings to old words and symbols. The creative aspect is that one moves in this type of behavior on the meta-level. We are not learning new signs or symbols, but new rules to control those symbols. This means that we are taking here a meta-standpoint in relation to the previous level of learning, the semantic conditioned reflexes. The whole semantic field of a branch of science or of a practical activity or of human culture as such is restructured when a new grammar replaces a previous one.

But what is exactly the creative character of this type of learning process? One could express this first in a negative way by stating that the three previous levels of learning as such do not go beyond a certain fixation. The Pavlov dog in every human being receives new stimuli as signs substituting hereditary ones. This substitution takes place on the basis of natural tendencies and is reinforced by punishments and rewards. Also on the second level such a type of purely conditioned behavior is evoked in the process whereby new responses are learned by trial and error.

Some authors, such as Karl Popper, explain the whole procedure of hypothesis building by such a system of trial and error, of conjectures and refutations. And in the wider field of human culture, more biologically oriented writers also affirm that the history of culture is a process of trial and error. But all this still leads to new, but nevertheless fixed, patterns of behavior. Only when the whole frame of reference, giving signification to the substitution of both stimuli and responses, is being changed, fixations are eliminated and therefore creativity is related to restructuring the whole semantic field.

In a more positive way creativity is not only envisaged as a breakthrough beyond a certain fixation. Often it is said to consist of a sudden flash of insight. But that is only a psychological description, a description, moreover, that often does not fit the facts, one that too easily gives rise to an irrationalistic and romantic conception of creativity. Creativity, on the contrary, is the constitution of a new type of rationality. It is not just an inexplicable type of insight; it is the highest achievement in human learning behavior. Therefore, a more positive formulation of creativity describes it as the knowledge of the problematic aspect of a network of symbolic actions, in sciences as well as in daily life. Knowing the problematic side of existing data implies the possibility of going beyond those data as such. In order to produce a new grammar it is necessary to go beyond the scope of the given information and the given structures. If one prefers a more psychologically tinged formulation, then it is better not to say that creativity is a flash of insight, but that it is a sensitivity to problems. Going beyond the given frame of reference is a sign of transcendence, or transcending what is presented to us.

Creativity is a normal human learning process consisting of saying "no" to an existing system of rules in order to renew the whole scope of already given solutions. Often a new structure or grammar is related to the old one in the sense that it gives an enlargement to the previous grammar. Many authors in the field of the history of sciences even say that every renewal consists of making the previous frame of reference into a special case, a borderline case, of a wider network. Euclidean geometry is then a special case of non-Euclidean geometries, Newtonian mechanics of Einsteinian theory, classical economical theories of Keynes' theory, etc. Others, among them Kuhn, are of the opinion that this type of structural revolution is not so rectilinear and that therefore even the same words or terms, for instance "mass" in Newtonian of Einsteinian mechanics, have a different meaning. Both standpoints can be reconciled in the sense that the creativity of learning a new grammar or structure remains a learning process and is to that extent still based on previous experience and previously accepted structures. Creativity is not a supernatural, miraculous procedure. On the other hand creativity does indeed imply an activity or restructuring which results in new meanings for old words and symbols. The sensitivity to problems is always bestowing a new dimension on previous networks of symbols.

This fourth level of learning processes consists of more than problem solving. More is required than just solving problems. One is in need of new rules for the solution of problems. One is often even in need of turning existing solutions into new problems. The accent is more laid on the "how" than on the "what." This process of creative learning does not mainly aim at questions about what things are, questions of explanation, but at questions about how one has to proceed, questions of strategy. One can call creativity a new insight, but this expression has then to lose its incidental character and must become an indication for a way of operating. Such a procedure of acquiring a new insight does not go without rules, but it constitutes, on the contrary, new rules.

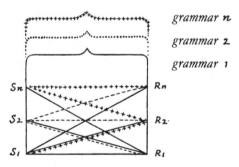

II. Creativity as construction and as response

This survey has so far had two aims: first, to present a comprehensive theory about human learning behavior, eliminating the reductionist's attempts to reduce such a behavior to its biological basis. Second, this survey has aimed at a more philosophical interpretation of human culture, an interpretation which does not begin from a spiritualistic viewpoint, but stresses on the contrary the analyzable, in this wider sense "behavioristic," aspect of all human culture. Complementary to this aspect the survey has tried to show how real creativity takes place such that the human central nervous system, exteriorized in the whole pattern of human behavior, manifests the possibility of transcending, surmounting every inventory of data. Creativity implies that human conduct does not have to take the data for granted. In other words, human behavior shows a dimension of transcendence which becomes visible in the immanence of the given dynamic structures of human action.

What has been formulated now as the interwovenness of transcendence and immanence in human conduct can also be reformulated as the two sides of human culture as such. The term "culture," against the background of human learning behavior, is then not to be taken as a noun, a substantive, but as a verb, the way in which humanity functions in history. There is a restlessness in human culture which speaks of the everlasting creative efforts to renew old situations, to reshape existing structures. This dimension of transcendence has to be taken account of in every explanation and description of human history and culture.

In the contemporary situation of cultural transition, a sharpened awareness is to be found of this dimension of creativity in human cultural behavior. An effort can be felt in this whole realm to become conscious of the structures, the grammars underlying the field of culture. This structural awareness often points to only indirectly present structures and terms like "depth-grammar," "hidden structures," "meta-research," and "metascience" express this awareness. Also in the field of artistic expression the structural impact is great. The artistic data are not dominating the structures any more but the structures are decisive as to the meaning of their content. Cultural anthropologists, general linguistics, philosophers, and historians of sciences reflect upon the changeability of structures; Lévi-Strauss, Kluckhohn, Chomsky, Foucault, Barthes, Koyré, and Kuhn try to find different solutions. Some of them point to deep structures that do not change, others on the contrary speak about "paradigms" and "epistemes" that are in constant process of change. We are not entering the field of this discussion here. The only thing we want to do is to draw attention to this intensified awareness of frames of reference that are decisive for the meaning and signification of the whole symbolic network of culture.

The meaning of culture is a historical one. It manifests changes in structural meaning. This goes far beyond merely semantic learning. It is a pragmatic process in the sense that the semantic meaning is related to those who are making use of symbols, be it in more theoretical reflection of sciences, or in more practical activities of fashion, political activities and social organization. Culture is, in this way, the sedimentation of collective human learning. The renewal of structures, the substitution of grammars, in thought and action, is the explosive kernel of culture that distinguishes human behavior from animal reflexes. This explosive center consists exactly of the focus of creativity: the sensitivity to the problematic. Human culture is always an encounter with new possibilities, with a partly unknown future. Creativity arises at the borderline between the existing and the not yet existing culture. The transcendence in the immanence of cultural structures is the pushing force of creative restructuring. The essence of human creative learning consists of the incongruity of any culture with the surrounding experience. The awareness of incongruity is the urge for the historical process.

It may be useful to try to put this in other words. Culture is the human approach to reality. From agriculture to the cult of the gods, from the whirling dance to the clear and cool formal systems of logic and mathematics, culture is an approach to realities that are never completely exhausted. Culture as activity, as verb, is the constructive effort to delineate its own incongruity. In this sense the reality approached by culture is not so much pointed to as a "what" but more as a "how." It is the way of operating, of procedure, that constitutes the references to reality, just by manifesting the incongruity of the cultural construction of reality.

In order to avoid misunderstanding it must be repeated that the matter of incongruity as structural motivation in culture is something very positive. The highest level of learning behavior is only possible as a restructuring on the basis of awareness of the problematic aspect of any given structure or grammar. Creativity is exactly this effort to go beyond what is given. Culture, as dynamic force, implies the history of constant renewal. Some cultures give the impression of being more static, but of course every culture goes beyond the static nature of hereditary-given animal instincts. In all cultures some urge towards wider horizons, new experiences, and mythical perspectives is to be found. The term "incongruity" therefore points to the surprising disposition of any human cultural behavior to renew the known by the awareness of the unknown. In daily life the solution of a problem can be given as soon as one succeeds in throwing new light on an old dilemma. In culture it is possible to cope with new situation when old answers are reformulated in a really new way and when traditional approaches are fundamentally questioned. For the individual learning behavior as well as for the progress of culture the question marks are more fundamental than the notes of exclamation.

It is necessary to elaborate this aspect of cultural activity in another direction so that the previous remarks receive more background. The issue of this part of this essay will be the relationship between culture and reality. To what extent is culture something real? Is it only an illusion? Is it, as H. Vaihinger interpreted it, only the construction of "as-if's," of fictions? They are misleading as to their real content, but at the same time indispensable for the biological preservation of our existence. Our answer will be that culture is not an activity "as-if," a behavior dominated by illusions, but, on the contrary, culture so much touches the real world around us that we could not even use words like "reality," or "natural world" if they were outside the scope of the constructive activities of human culture.

First, an explanation has to be given for the repeated usage of the two words "culture" and "construction." The thesis is that culture is constructive, this term used in opposition with mirroring or reflecting. A more passive conception of culture regards it as a kind of human looking-glass, reflecting everything that is going on outside man: culture mirrors nature, in daily experience, sciences and the arts; it is often also seen as mirroring supra-nature, in myth, and metaphysics. But in all these looking glass functions culture is reproducing. The main feature of cultural activity, however, is not reproductive but to some extent productive. We saw in the hierarchy of learning processes that perhaps the lowest level, that of conditioned reflexes, is of a more passive character: the animal has to accept the substitutes associated with the original stimulus. But already on the second level, where the response is being replaced by a new one, the learning behavior becomes more active. Here, we said, a central nervous system is indispensable. And when in the scale of living beings this central nervous system becomes more complex, as the cortex becomes functional, then the activity of the learning behavior becomes more apparent, in semantic reflexes as well as in the possibility of substituting the whole grammar of a network of symbolic action. The collective learning process which we call "culture" is thus marked by a high degree of activity and is in this sense more productive than reproductive.

Symbols replace stimuli as well as responses; restructuring is the activity of replacing a whole grammar of a system of symbols. Cultural behavior is the process of symbolizing the surrounding world as well as the own behavior. It is a process which works in two directions: it implies the transformation of data, the transformation of nature, and it contributes to the identity of the person or group that is the agent of this symbolizing activity. A certain culture gives at the same time access to the symbolically disclosed world as to the identity of men and peoples involved in this specific culture.

There is a tendency to regard culture as the subjective superstructure based on the substructure of objective data. These data are, according to some authors, the verifiable data and facts of natural sciences; according to others they are the material conditions for society; and again others seek this stratum of objectivity in the biological nature of man. Culture is then regarded as a subjective impression or a kind of collective ideology. To culture belong the values and world views of a group of human beings, and to objective nature belong the states of affairs. Values belong to a kind of emotional and subjective language and to an objective universe of discourse. This dichotomy between subjective values and evaluative language on the one side and objective facts and constative language on the other side has to be rejected for the simple reason that the "facts," the states of affairs, the natural sciences, the material substructure etcetera are themselves the result of a human process of symbolic activity. The various trends in contemporary philosophy and sciences of culture that have been mentioned before, like structuralism, phenomenology, and the new philosophies of science, make clear that there are no facts outside a theoretical conception, no biological nature outside a certain cultural attitude, no substructure outside the frame of reference pointing to superstructures. Fact finding, verification and falsification, and conceptualization of nature are also results of an active process of human learning and the outcome of the human ability to operate with and through symbols. Culture is exactly this interrelatedness of facts and values, and the various scientific disciplines use different criteria for "objectivity," "facts," and even for the explanation of evaluative behavior and language.

A different approach may explain further how this dualism between nature and culture, objective reality and subjective response, has to be overcome. A variant of the mirror theory of culture is the "white stain" theory. This means that reality is regarded as a map; but the field of reality being not fully mapped out, there remain white regions, white stains on the map. Human knowledge, the sciences, and in general all cultural activities are regarded as a process by which the white stain becomes smaller and smaller, although perhaps endlessly. In a way this analogy has as presupposition the separation of man and world, of the knower and the known, of cultural behavior and reality. But the situation is much more complicated. Culture is not a linear activity of discovering the world. It is not a simple ongoing process by which reality gradually becomes better known and better controllable. The most characteristic feature of culture is its pluralistic character. There are many cultural approaches at the same time. There are also many cultural strategies that succeed one another in the history of the same culture. Culture as verb is not rectilinear but is more zigzagging. Or better still, culture is not the process of continuous discovery, it is the ongoing process of restructuring. In art as well as in sciences, in the continuous self-reinterpretation of various religions as well as in the meandering flow of economic order, reality shows always new aspects and structures. The existing map is not filled up, but rewritten, again and again, in new structures and new symbols.

The consequences of all this are far-reaching, the main conclusion being that reality is not something to be found outside the field of that learning process that we call "culture." Culture is not a subjective, or even subjectivistic, approach to a static, objective reality. Culture and reality are interrelated to such a degree that the restructuring history of culture belongs to reality itself. Culture is not a discovery, often in a subjective way, from the outside. It is the disclosure of reality from the inside. Culture is not a kind of flight away from an objective reality. It is not a merely subjective response to the invariable stimulus of a surrounding nature. Culture is the activity of engaging oneself in what is real and valuable. Or in the formulation already given: it is not a discovery but a disclosure. And the same seen from the pole of reality: reality is not like a stone hidden in the ground waiting to be excavated, like a silent immutable entity; reality is the appeal for restless activity. The essence of reality is that it has to be expressed, put into words. And it is this feature of reality itself which we call human culture. Culture as the activity of construction is as much a feature of reality itself as of human behavior.

Perhaps all this sounds more poetic than convincing. It is worthwhile therefore to go back to the more exact interpretation of culture as a human learning process. In all learning behavior there is something of trial and error. Even the Pavlov dog sometimes gives no response at all to a new stimulus offered to him in a situation that is nevertheless of vital importance because of the reward or punishment linked to it. In the natural history of evolution animals develop higher degrees of nervous systems, intelligent behavior, and learning processes on the condition that their responses are adequate and fit to the surrounding world. This adequacy is indispensable for the success of a learning process. Even the whole organic and neurophysiological constitution of a living being is the outcome of the adequacy of such an organism and its behavior. The same is valid for the system of human behavior, indicated by the term "culture." Cultures come into existence and cultures perish. But cultural behavior as such implies a kind of adequacy in the interaction of man and world.

"Adequate behavior" does not mean that the behavior mirrors

its surrounding world. A hammer need not reflect the nail in order to be a useful instrument. Symbols do not mirror the outside reality in order to become instruments for the disclosure of that reality. Symbols, and the more a whole grammar of symbols, are more productive than reproductive, more constructing than mirroring. Man is as much a biological as a cultural being. His biological constitution is interwoven with his cultural behavior. The human constitution matches in general the surrounding reality. Of course, many mistakes, failures, and misinterpretations mark the historical path of mankind, even to today. But on a more abstract level one could say that symbolic and cultural behavior could only come into existence as more or less adequate responses to reality. The remarkable feature of this adequacy is that it matches reality not by reflecting and mirroring it, but by a more active disclosure of reality. Symbolic transformation appears to be the human response to surrounding reality: it is an intrinsic feature of reality that cultural transformation must do justice to it.

Perhaps we must begin to use the terms "real" and "reality" in a new way. The meaning of these words is mainly copied from the ways in which we use material objects and other stable things. But like the term "culture" the term "reality" also has to be taken more as a verb than as a noun. Reality is more a dynamic task than a static datum. It is not a substantial substratum to be found outside and before the scope of human symbolic interpretation. On the contrary, reality exists only as a kind of story which has to be told by human culture. Telling a story is always a symbolic interpretation. The way it is told can be reliable or unreliable. The dimension of evaluation and ethical responsibility is essential for the human response to reality. And in this perspective one might even say that this dimension of ethical responsibility is essential for reality itself, as it demands such a human response and as that dimension is indispensable for the adequacy in the relationship between man and world.

One aspect of human learning is that man can be conditioned in such a process. This is especially apparent on the lower levels of learning. Here the possibility of manipulation of man and society is to be found. But here also one meets with the tendency of man himself to obtain certainty and fixation of behavior. The way in which the Pavlov dog is learning to distinguish between a circle and an ellipse can be very attractive. If the animal performs an adapted behavior without the necessity of interpreting the signs that are presented to him, that means that the signs need not to be taken as symbols (the X-ray photo of sick lungs acts only as signal and does not for the dog refer to the phenomenon of illness). There is a human tendency to seek the security of this type of animal conditioned behavior, even on the higher levels of learning. Philosophers sometimes dream of a perfect techno-logical society with a well-balanced system of psychic and social conditioned reflexes. The higher the level of learning, the more insecurity and risk are built into the process of interaction of man and world. But even on the highest level one tries to obtain fixations and almost vital security. An example is metaphysical escapism, in which the total system of reality is conceived of as substantial entities ruled by fixed laws. It is an escapism because it tries to get away from the openness and changeability of seman-tic (symbols) and structural (grammars) learning. It is metaphysical because it strives for fixation outside the realm of empirical culture and historical changes. It speaks about Being, the determinism of eternal laws and the fixation of events in substantives. In a way, this escapism could be regarded as sublimation on the level of culture of the Pavlov type of learning process.

These phenomena in human conduct are also of practical importance. Every type of instruction and education has to do with the effort to learn creativity, to acquire a creative behavior and flexibility of action. Many systems of education and various books of instruction are a kind of metaphysical escapism. They point to fixations outside the field of constant renewal and continuous learning. Instead of this, what is needed is a type of creative learning. The requirements for this are, first, the stimu-lation of the awareness of frames of reference. Often these frames are hidden and at the same time, in a way, so obvious that they hardly enter into consciousness and awareness. Second, the element of transcendence is necessary, which means that one has to become aware of the incongruity of every structure with regard to the dynamic reality to which it responds. From the sensitivity to the problematic, as intrinsically belonging to every comprehensive structure or grammar, arises a third requirement: positive flexibility, the openness for new solutions.

As has been said before, all this is not restricted to psychological and didactical advice. It has an ontological background, even a metaphysical one, but it is not the metaphysics of escapism, rather of the interaction of transcendence and immanence. Or to put it in more simple terms, it has to do with a certain conception of reality itself. What do we mean by saying that something is really true, or really good, or effective? The word "real" indicates that a certain human opinion or item of behavior is flexible, giving new answers to older problems, being at the same time adequate, as it apparently does justice to the requirements of a situation. It points to a certain human activity that is in an authentic way engaged in reality, just because it is restructuring that reality towards its further disclosure. Or to say it in a still shorter formula: a human being becomes creative when he concerns himself with reality in a very active and open way. In this sense one could represent the ontological formulation that reality itself exists in a necessary relationship with human symbolic transformations. Once again, by way of summary, the story told about culture and about human learning behavior is not so much a description of the psychology of learning as an explanation about the dynamic structure of reality.

A last remark has to be added to this. The term "symbol" has been used many times. But we must not try to fly now into metaphysics of symbols as a new type of static entities. "Symbol" is a task-word. No symbol is without the interpretative activity of man. P. Ricoeur stresses the fact that every symbol has to be re-interpreted throughout history. He speaks about "hermeneutics" as the effort to resume the meaning of a symbol in always new interpretations. This does not exclude the continuity in the inter-pretation, for instance of the great religious documents and histories. It means on the contrary that this continuity does not exist as simple repetition but by expressing the same significance in a new way. One could even say that it remains the same signifi-cation only when the renewal of interpretation is brought about. Culture as a structural field of symbols is doing justice to reality, reality not only in the field of natural sciences, but certainly also with divine reality, when creative renewal expresses the positive awareness of the problematic, that is the dynamic, character of every human approach. In this way the transcendent dimension is indispensable for seeing with clear eyes the immanence in which we find ourselves.

III. Capitulation as human strategy

Strategy is part of human conduct. The reality surrounding man demands a flexible approach. Reality is not a collection of objects but a series of events, like A.N. Whitehead, H. Bergson and others have tried to show. One could say that reality itself is more like a text than the paper on which the text has been printed. The strategy is the effort to find access to the text of the world. This world is a historical one, with a changing nature, disclosed in always new interpretations, often requiring a new grammar, as evidenced from Ptolemy to Copernicus, from Newton to Einstein, from Linnaeus to Darwin. This world is the world of human culture with its various approaches and its rewording of the old text of reality.

The term "strategy" derives from military vocabulary. It also plays a role in the theory of games, which in modern times has been developed as a partly mathematical theory, for instance by O. Morgenstern. Here it gives insight into situations where people try to realize a goal, taking into account the eventual counter-measures of other groups or individuals. It is well known in chess how various strategies can be used. Someone who knows little about chess sees only the various moves on the chessboard. But the more qualified spectator will observe that the chessmaster is not simply executing a new move but that he is changing his strategy: he jumps from a previous strategy to an alternative one. A strategy could be defined as a certain rule that regulates a series of separate moves or operations. The moves, the operations, the various human actions are directly visible, but the strategy is only indirectly observable. It acts as a kind of hidden theorem from which the separate operations have to be deduced.

With the investigation of human strategies we find ourselves on a level of highly symbolic behavior. A strategy implies the knowledge or at any rate the capacity for knowledge, of other strategies. In this sense strategic behavior belongs to the highest level of human learning processes as it presupposes the possibility to change the total grammar of a system of symbols. The same move on a chessboard can be derived from two different strategies. Strategy is one of those concepts that tries to cope with the hidden dimension in every human behavior. It is not hidden in the metaphysical sense that it is as such outside the scope of direct and rational analysis. It is hidden only in the sense that it is indirectly given; it is not empirically present, but is at the same time indispensable for the explanation of what is directly given.

The original use of the term "strategy" implies the effort to obtain a certain goal as well as the resistance of an adversary. In the military usage of this word both meanings, the goal or objective and that of the adversary, are well-defined. In a wider and more modern usage this is not so much the case. In the older conception of strategy the "what" was more important than the "how." In a newer conception the "what" cannot be neglected, but its definition depends partly on the "how" of the strategy. In management, for instance, terms like "management by objectives" and "corporate strategy," as well as "long-term planning," are often used. But in many cases, especially in large-scale situations, the definition of the objectives often has to be reformulated under pressure of the strategy. Of course, this strategy is also flexible, and so a kind of interaction takes place between the reality of the goals and the strategy of the behavior. Both are no longer things standing by themselves, but rather are constituted in a process. The opponent or adversary also becomes less definable. In a market situation the adversary can be a competing firm or economic fluctuation or activities of action groups. But according to different strategies the type of adversary can change.

A still wider use of the term "strategy" can be made if applied to human culture. The whole of human history can be seen not as a mere passive response to the forces of nature, life and death, but as an often unconscious strategy by which structures and grammars are changed, and by which man tries to cope with the often unseen powers of fertility and disaster, happiness and misery, good and bad consciousness. One has often applied to the process of culture the biological criterion of the effort of survival. But in the symbolic network of a culture, sacrifice, awe, compassion, utopian dreams, and mystical visions also play a tremendous role.

The strategy of a culture cannot be reduced to the fixed vital instincts. The objectives of every strategy can be different, varying from power, rational bureaucracy and economic welfare to immortality, the love of mankind and religious redemption. Here also the adversary as well as the objectives are interrelated with the strategy itself. The striving after immortality, to give an example, does not dictate one special strategy and can even, according to the strategy, be qualified as a lofty one, for instance in the profound respect for the unselfishness of the eternal Ideas or Forms in Plato's philosophy. But it can also appear to be a very egocentric attitude, or a fixation of historical responsibilities. The awareness that various moves and steps are regulated by a strategy and that this strategy is the "how," the way in which the various actions are performed, can have such an influence on a previously established goal that this changes under pressure of the responsive, even responsible, character of a strategy. Especially in times of rapid cultural change, the awareness of a strategy is necessary.

The adversary of the strategy of culture is still more symbolically interpreted and encountered according to the strategy. What are the powers with which humanity, or a group of people, are confronted? Are they the powers that govern the means of production, or, like in a mythical society, the forces dominating fertility and sexuality, or are they the hidden persuaders, perhaps archetypical forces in the depth of our souls, or the divine influences traceable in exceptional as well as in daily situations? In a way human history could be compared with a continuously changing chessboard. Man develops his strategies and gains in this game his own identity. The adversary is only visible through the symbolic interpretations. Religion, sciences, the arts, social development, daily experiences, all this is to be found in this field of symbols. Sometimes man pretends that he himself is his own opponent; sometimes he discloses in the powers of reality an enemy, sometimes a friend. But it is one field of phenomena and it is useless to separate scientific and objective thinking from this fundamental strategic effort, the frame of reference of all human interpretation. Even behind the deductive-nomological skeletons of scientific explanation a certain policy is to be found, which is different according to culture and historical period. Science is one of the media which copes with reality, and it forms a part of the organism of human culture. The awareness of strategies gives clarification about the unity of culture as well as about the directives and perspectives of man's participation in the story of reality with its dimension of powers to be identified.

There are various strategies and every strategy can be realized in a variety of ways. A strategy is always the main rule, the regulating grammar, of human action and thought. This implies two things. On the one hand a strategy has a very open character. Feeling and intuition can play an important role. Great generals, but also pioneering artists, yes, even children in their ingenuity of inventing new plays or people finding new approaches in the daily course of events, they all develop new strategies. But this does not mean that a strategy belongs to the merely emotional, non-cognitive part of the human possibilities. A new strategy is a change in grammar, the creative, but in a way normal, part of any human conduct. So it can be analyzed, it has a structural aspect. You can even make a kind of blueprint of a strategy, be it only schemes and models of the more complex process going on. In other words: a strategy can belong to the psychic field of intuition, when we regard it from the standpoint of the individual or group concerned; we may call this the biographical point of view. But at the same time a theoretical analysis remains possible for the outsider, the more objective spectator; we may call this the methodological point of view. Or to formulate it in more difficult but classical terms, derived from distinctions made by Kant: a strategy has as well an aspect of *quaestio facti* (factual, empirical investigation) as an aspect of *quaestio iuris* (transcendental-logical investigation). This results in the conclusion that the investigation of strategies needs the interplay between empirical sciences and logical-philosophical research. Or, to put it in the terminology of contemporary issues: the context of discovery and the context of justification can be distinguished only in an artificial way.

The term "capitulation" belongs mainly to the military usage of words connected with the term "strategy." It means, literally, the drafting of chapters or capita regulating the ways in which the surrender of an army takes place. In a wider sense we understand by "capitulation" that one renounces a position. The difference with the word "surrender" is that in the term "capitulation" an aspect of explicit consideration, almost of calculation and computation, is present. Therefore, it fits into the vocabulary of strategic action provided that the creative and noncognitive aspects remain in the background of the usage of the term "capitulation."

As the term "strategy" is connected with the presence of an adversary, so the term "capitulation" is related to superior forces of the opponent, and in general, to something like *force majeur.* It is understandable that in daily language "capitulation" is mostly associated with a passive surrender, with abandonment. But as has been said before, there is something of calculation, of deliberate action in a capitulation. If capitulation indeed constitutes a part of a strategy, then it has an element of anticipation in it. In the framework of the present discussion of culture as a human strategy, conscious or unconscious, we must stress this active aspect of the meaning of the word "capitulation." This does not exclude that a capitulation frequently takes place when no way out can be found in a difficult situation. In that case, the capitulation is the breaking off of a strategy and could hardly be considered as an integrating part of that strategy. There is also, however, a more deliberate form of capitulation. We call this "active capitulation," and that term indicates that a certain objective is intended by such a part of the strategy.

These preliminary remarks on "capitulation" must be elaborated in order to get further clarification. A simple example can serve as introduction. When "strategy" indicates mostly a long policy — although "strategy" has a much wider meaning, acting in the whole field of cultural structures — the term "tactics" is comparable to a short-term policy. Now it is clear that in the field of tactics a capitulation can have a very positive and effective role. To abandon a certain stronghold can be the best means to arrive at a final victory. To abandon intentionally a chessman can improve the strategic situation. All these analogies, however, do not go far enough. Is it conceivable that one resigns in a game of chess and that this still forms a part of a positive strategy? That is imaginable, but then the scope of the strategy becomes much wider, even very different. It might be possible that during the game one of the chess players transforms his strategy, and the objectives of his game, to a scope far beyond the boundaries of the rules of chess. Let us imagine that our chess player is a good psychologist and that he discovers that his opponent finds himself, in his private life, in an extremely difficult situation. His opponent has perhaps the feeling that he cannot anymore face the problems of his own life. Our psychologically gifted chess player might even perceive that the actual chess game is the symbolic field on which his opponent projects, unconsciously, the situation of his daily life. At that very moment, the strategy of our chess player gets a wider and new scope. His aim is now to encourage his opponent and to stimulate him in order to overcome, in this very game of chess, his obstacle within the restricted, but symbolic, limits of this game. His capitulation is then a very deliberate one, indispensable in realizing his new strategy.

This is only a story and an example. One might be inclined to dispute the whole idea of a capitulation, because of the fact that losing this game of chess means obtaining the goal of helping his opponent. But we must not forget that this goal is only meaningful within the structure of the new strategy. If one fixes the given situation that is, the situation defined by the rules of chess then the new goal is meaningless, even nonsensical. One can even imagine that a qualified chess player, who learns the whole story about this therapeutic approach, is nevertheless of the opinion that the chess player has acted more as psychologist or psychiatrist than as chess player and that he is, indeed, a bad chess player, who lost the game.

Now this story becomes more exemplary, a kind of parable. When somebody remains within the limits of a given strategy and an existing grammar then all intentions and objectives, defined by a strategy that transcends the previous one, are meaningless. When we confine ourselves to Euclidean geometry, the statement that the three angles of a triangle might be more or less than $180°$ is nonsensical. When we restrict our scope to the fields covered by natural or even social sciences, or by technological planning then many ideals are illusory. It is not possible, in a very abstract manner, to state that a capitulation is not a real ideal, as the new strategy can realize its objectives, but whether the new strategy is as such a strategy or only chimera. Our modern society cannot be helped by general advice to develop new strategies. The only real and concrete aid to contemporary society consists of an ethical decision about the real or unreal character of a new strategy. And to make such a decision implies the realization of effective capitulations in the field of an existing strategy.

A capitulation remains a capitulation, even when it is an active contribution fostering a new strategy. It receives its meaning as capitulation from the grammar to which it belongs. And within the limits of that grammar it implies the end of a strategy. Such an end can be final, or it can be the transition to a new strategy. But, at any rate, the new strategy requires a new way of expression, a new structure that holds a network of symbols together. And a new way of expression cannot be understood within the rules of the old system. But is there no continuity whatsoever? Are there no means of translation? Yes there are, but only in an indirect way. Perhaps it is exactly the capitulation itself, or at any rate the way in which it is performed, that presents the best indication of what is aimed at in the new strategy. It has been said before that such a capitulation is an anticipation. It belongs to the renewal of a strategy that this anticipatory character of a capitulation is shown in the whole context in which it takes place. In our parable or story this requires the effort to look beyond the edge of the chessboard, to look beyond the restrictions of scientific and economic models, to look beyond the boundaries of many seemingly unescapable situations, in the lives of people. This effort to discover transcendence in the immanence is the ethical motivation of the renewal of a culture. The qualification "ethical" signifies for the moment only that a decision is required as to the real or authentic character of a new strategy, a decision that cannot make use of the accepted and merely cognitive criteria of the existing strategy.

A capitulation as an event isolated from the context of a strategy cannot achieve a renewal. The ethical decision has to be taken within the existing strategy and in anticipation of a new strategy. It has just been remarked that the ethical aspect is related to the element of decision or, in other words, to the freedom to be taken in regard to the existing rules. But complementary to this the cognitive element, the requirement of analysis, must also be put forward. A strategic capitulation implies, as has been mentioned earlier, a kind of calculation. The term "capitulation" has a somewhat abstract and distant meaning. This is not meant in a negative sense. But a good capitulation demands insight into the whole context of the situation. A negative capitulation is just losing the game. A positive capitulation tries to go beyond the boundaries of the game. But then one must know the existing grammar, the rules of the actual game, in all details, with its limitations as well as with its possibilities.

A practical example may, again, help. Some years ago Jan Pallach, a student in Czechoslovakia burnt himself as a protest against the political situation. It was a self-sacrifice, a capitulation seen from the structures of power that were present at that moment. But Jan Pallach certainly hoped to achieve a fundamental change. And in the political history this hope came true, though not for a very long period. The act of Jan Pallach was a kind of trigger mechanism, which sparked off far-reaching political change. Pallach was not the only one to do this. Various students in Czechoslovakia burnt themselves also. Some years later the same phenomenon occurred in France. In Vietnam a Buddhist monk burnt himself with important political results. Various other monks, who did the same, did not effect any results whatsoever. Why in one case such a capitulation brought about such a direct effect whereas in so many other cases results failed to materialize? The answer is that Pallach and the first Buddhist monk did not act only as individuals and merely on the first impulse. No, they responded to the whole context of a political situation. They analyzed, partly intuitively but also on the basis of much information, the existing grammar of that situation. They perceived the weak spots in those power structures and the ethical gaps in that political situation. A positive capitulation is a capitulation that gets its meaning from an existing structure and that gets its anticipatory force from the signification of that capitulation in the light of an eventual new strategy. Knowledge of structures is necessary to be creative. Analysis of the context of a society is indispensable for individual action. The ethical impact of such a capitulation is supported by the well-founded investigation of the rules of an existing strategy. A positive capitulation is the translation in an immanent context of transcendent possibilities.

I have stated that the highest level of human learning must imply the awareness of the incongruity of every structure as to the dynamic reality to which it responds. A short program has been given in which reality itself evokes the structural response as well as the sense of incongruity from the side of human culture. In reality itself the transcendence and immanence are interwoven in such a way as to bring about the learning process of human history. Now all this has been formulated in more concrete examples. The sketched ontology is about a reality which demands human creative and renewing action. It is a reality which must not be thought of as a distant object or a naturally given substratum, but as a dynamic appeal, even as a history in which human action is involved. This type of ontology might be called a functional ontology and this simply means that reality is interrelated with human action and human thought. This relation to man is not accidental but essential or structural. Reality intrinsically forces itself on human learning behavior, that is on human history. Human learning behavior is only successful when it does justice to that reality, which we called the adequacy of human symbolic behavior. This term must not be taken in the old sense, such as when it is used in one of the classical definitions of truth: the adequacy of the object with the representation of that object in human understanding, *adequatio rei et intellectus*. In such a conception one starts from a separation between the subject and the object of knowledge. But the adequacy of symbolic behavior means that the interrelatedness constitutes the starting point. Or in other words: the symbolic transformation of data by man in the whole of his cultural behavior does not mean that he changes a distant and static reality, moulding it according to his own subjective ideas: it means on the contrary that a symbolic transformation that does justice to reality accords with the inner structure of reality that demands such a continuous and dynamic interpretation. A good grammar of symbols, in science, ethics, religion, social action, games, sport, etc., is not an interpretation from the outside, but the involvement of man in the story of reality itself.

Now it may become clear, on the basis of this rather succinct outline of such a dynamic ontology, that reality does not have to be found behind daily phenomena. It is not a substratum or a hidden substance or a static object outside the field of human interpretations. It has to be sought in the field directly given to us, in daily experience and in the phenomena around us. We explained how the adequacy of a system of symbols, a pattern of cultural behavior, goes hand in hand with the awareness of the incongruity, the sensitivity to problems. Now it is possible to explain this in terms of other formulations as well. The term "creative capitulation" is used to indicate this double aspect of adequacy and incongruity. In practical action adequacy manifests itself as part of human creative behavior; and the incongruity takes a more concrete shape as the possibility of a capitulation that is not merely a negative one. If it is indeed true that the reality around us has such an appealing and dynamic structure that it demands the symbolic effort of man, then reality is a verb, a verb that has to be conjugated in the events of social, political, and religious human history. This implies that the structures of reality can be reformulated in more human, even personalistic, terms. A philosophical analysis brings reality to the fore in the personal attitudes of individuals, of groups and of whole cultures. Man is playing his game of chess and he is forced to follow a strategy of symbolic interpretation. His opponent, the forces humanity is being confronted with throughout its history, can only be discovered through the symbols of religion, ethical responsibility, and human thought. And the objectives of his strategy are not unequivocal. The objectives can even pass through a metamorphosis during the transformations of the strategy itself. Symbolic behavior implies always a certain ambiguity. But just for that reason it demands not passivity but human activity. Man is an "animal symbolicum" as E. Cassirer said. But this does not mean, as he thought, that the human mind could gradually master reality, reducing it to a subtle network of artistic, mythical and above all scientific symbols. The ambiguity remains. It is essential that man cannot always reduce reality to a symbolic transformation which turns out to be reliable or unreliable. The symbolism is not so much the expression of the human mind, as Cassirer thought, but of the adversary, of reality that demands to be wrestled with. As in the story of Jacob wrestling with a supernatural power in the dead of night, the creative capitulation points to possibilities beyond human behavior. The transformation of an existing grammar is more than merely a linguistic process. It is human creativity as learning process, sensitivity to the problem of transcending given situations, human culture as verb with infinite conjugation.

University of Leyden

References

Gaston Bachelard, *La Philosophie du Non* (Paris: Presses Universitaires de France, 1949).

Henri Bergson, *Les Deux Sources de la Morale et de la Religion* (Paris: Presses Universitaires de France, 1932).

Ludwig von Bertalanffy, *General System Theory* (New York: Brazillier, 1968).

William I. Beveridge, *The Art of Scientific Investigation* (New York: Random House, 1950).

Donald I. Broadbent, *Behaviour* (London: Methuen, 1961).

Ernst Cassirer, *An Essay on Man* (New Haven: Yale University Press, 1944).

Noam Chomsky, *Language and Mind* (New York: Harcourt, 1968).

Charles H. Elliot, *The Shape of Intelligence: The Evolution of the Brain* (London: Allen and Unwin, 1970).

Gerhard Frey, *Sprache, Ausdruck des Bewusstseins* (Stuttgart: Kohlhammer, 1965).

Donald O. Hebb, *The Organization of Behavior* (New York, London: Wiley, 1949).

Ernest R. Hilgard, G.H. Bower, *Theories of Learning*, rev. ed. (New York: Appleton-Century-Crofts, 1965).

Susanne K. Langer, *Philosophy in a New Key* (Cambridge, Mass.: Harvard University Press, 1942).

A.R. Luria, *Higher Cortical Functions in Man* (London: Tavistock, 1966).

Maurice Merleau-Ponty, *La Structure du Comportement* (Paris: Presses Universitaires de France, 1949).

Orval H. Mowrer, *Learning Theory and the Symbolic Processes* (New York: Wiley, 1960).

G. Murphy, *Human Potentialities* (London: Allen and Unwin, 1957).

David Novitz, *Pictures and Their Use in Communication* (The Hague: Martinus Nijhoff, 1977).

I.P. Pavlov, *Conditioned Reflexes* (London: Oxford University Press, 1927).

C.A. van Peursen, *The Strategy of Culture* (Amsterdam, Oxford, New York, 1974).

C.A. van Peursen, *Wirklichkeit als Ereignis* (Freiburg, München: K. Alber, 1971).

Michael Polanyi, *Personal Knowledge* (Chicago: University of Chicago Press, 1962).

Karl Popper, *The Logic of Scientific Discovery* (New York: Basic Books, 1969).

Paul Ricoeur, *The Conflict of Interpretations* (Evanston: Northwestern University Press, 1974).

Paul Ricoeur, *La Métaphore Vive* (Paris: Ed. du Seuil, 1975).

W. Schapp, *Philosophie der Geschichten* (Leer: Rautenberg, 1959).

John Searle, *Speech Acts* (Cambridge: Cambridge University Press, 1969).

B.F. Skinner, *Verbal Behavior* (New York: Appleton-Century-Crofts, 1957).

Arthur A. Staats, Carolyn K. Staats, *Complex Human Behavior: A Systematic Extension of Learning* (New York: Holt, Rinehart, and Winston, 1963).

William H. Thorpe, *Biology and the Nature of Man* (London: Oxford University Press, 1963).

CREATING AND BECOMING*

MICHAEL KRAUSZ

I

In his article, "Creative Product and Creative Process in Science and Art," Larry Briskman argues that any theory of creativity must hold creative products as logically prior to creative processes. His general idea is that one cannot identify what a creative process is unless one begins with its results. The argumentative burden, then, falls on answering the question, What is a creative product?

This product-centered way of looking at the matter is consonant with the approach taken by Briskman's mentor, Karl Popper. Popper argues that all that can be said of interest about creativity in both science and art remains in the realm of the rationally reconstructable. A history of science is illuminating to the extent that it is capable of reflecting the logical or objective clashes between rival theories, reported observation statements, and the resolutions proposed by scientists. Similarly, Popper urges that the history of art should be understood in terms of objective problem situations that are more or less resolved by art works. This approach is taken up by Ernest Gombrich who, in his *Art and Illusion,*[1] shows how the introduction of perspective in Renaissance painting constituted a new visual solution to the objective problem of representing spatial dimension.

This product-centered viewpoint shuns biography as a mode of explanation of art objects, and in this respect there is a close affinity with the so-called New Critics, who hold that explanation or criticism of a work of art by biography constitutes the now famous "genetic fallacy." W.K. Wimsatt and Monroe C. Beardsley, for example, maintain that a work must be understood and assessed according to its internal coherence and force, and that must not be confused with the artist's psychological state or other contingent conditions of creation. The product-centered view follows the sharp

D. Dutton and M. Krausz, eds., The Concept of Creativity in Science and Art, pp. 187-200.

separation between the context of discovery (or creation) and the context of justification introduced by the logical positivists and refined by the Popperian idea of "rational reconstruction." These product-centered thinkers hold that the personal or subjective states of the creator are irrelevant in the understanding or assessment of works of science and art. They uphold an "objectivist" view of cultural entities, a view reiterated in Karl Popper's last three books, *Objective Knowledge,*[2] *Unended Quest,*[3] and *The Self and Its Brain.*[4] That view sharply separates subjective, mental, or feeling states on the one hand, and objective, logically reconstructable products on the other hand.

What, then, separates creative from non-creative products? Briskman holds that a product is creative if it breaks from its tradition, and by "tradition" he means a complex of background knowledge, problem situations, modes of analysis, puzzle solving, and the like. One is tempted here to invoke something like T.S. Kuhn's notion of a "paradigm" to explain Briskman's notion of "tradition" were it not for the controversy between Popperians and Kuhnians just on the issue of whether *any* new theory is tradition-breaking. That is, the Pooper-Kuhn debate turns on the question, Is scientific revolution perpetual or periodic?

Briskman does not provide a sufficiently clear account of what constitutes a "tradition" to be able to settle whether a given work is tradition-breaking. While some of the work of Beethoven may be said to be tradition-breaking in its departure from the rules of the classical sonata form, it is unclear how one could make a strong case for Bach's having broken an already established compositional model. We would more naturally think of Bach as having consummately refined a Baroque tradition rather than having broken it. Surely Briskman would not wish to invoke an ad hoc argument to the effect that consummatry is a kind of tradition-breaking, that is, the tradition of pedantry.

Most of the object-centered authors I have cited draw upon the history and epistemology of science as their basis for a general theory of creativity. It is not my concern to correct this by arguing for a general theory of creativity based on the history and epistemology of art instead, though one might try to do this. The reason that I shy away from doing so is that I am not convinced that the histories and epistemologies of science and art are, after all, symmetrical; nor, for that reason, that one might be able to offer a general theory of creativity that covers both realms. I am concerned here with creativity in art, and I am interested in the product-centered view to the extent that it is eligible within art.

II

I suggest that there is a systematic ambiguity in the notion of a "work of art," that this phrase is, as W.B. Gallie puts it, "essentially contested," that no clear and enduring set of conditions can be offered by way of a general characterization of it. This is so because its intensional construal may shift from period to period.

It is for this reason that I am most sympathetic with Nelson Goodman's suggestion that we replace the question, What is Art? with the question, When is art? Now, if the notion of a "work of art" is systematically illusive in the way I suggest, then any general theory of creativity which uses a general and univocal notion of a work of art as a type of creative product must be called into question. This is the basis of my unease with the position of Briskman and the other object-centered theorists.

Consider the case of Marcel Duchamp. In her book, *Conceptual Art*, Ursula Meyer says, "Duchamp rejected the myth of the precious and stylish *objet d'art*, a commodity for the benefit of museums and status seekers. His interest turned from the tradition of painting to the challenges of invention."[5] Duchamp's circumscription of concepts rather than precious objects as works of art was at once critical of the history of art and, at the same time, continuous with it. Again, Meyer says, "'The detachment of art's energy from the craft of tedious object production has further implications,' said Robert Morris. 'This reclamation of process refocusses art as an energy driving to change perception.'"[6]

The plasticity of the notion "work of art" as exemplified in conceptual art opens the way to regarding more than strictly static physical items as works of art.

III

In a different context Popper suggests that we entertain programs for ourselves, life programs, if you will. All these programs or schematic maps are theories about ourselves, and they help situate ourselves to ourselves. Popper says,

> This model or map, I suggest, with our own position marked on it, is part of our ordinary consciousness of self. It normally exists in the form of vague dispositions or programmes; but we can focus our attention upon it whenever we wish, whereupon it may become more elaborate and precise. This map or model is one of a great number of conjectural *theories* about the world which we hold and which we almost constantly call to our aid, as we go along and as we develop, specify, and realize, the programme and the timetable of the actions in which we are engaged.[7]

I suggest that the aims of what Popper designates as a program, or of what I prefer to call a personal program, may be regarded as a work of art. I shall provide a personal program of my own, taking license in doing so from Marcel Duchamp and the conceptualists as well as the related view of American Action Painters.

I am a painter as well as a philosopher and I am interested in the question: What is the relation between my self-development and my artistic development? This question arises not as an afterthought or as ancillary to my activity of making paintings. Rather, it is part and parcel of my understanding of what I think I am doing when I do art. If you like, my self-development and my production of art objects are integrally connected in my personal program. Of course, there are many artists who do not connect these two things.

The art *work* that I set for myself in my personal program may be stated as follows: my painting is instrumental to my self-development, and my self-development is instrumental to my painting. I value both painting and self-development as ends. That each is instrumental for the other is felicitous, but even if this were not so I should still value both as ends. By suggesting that artistic production leads to self-development I am trying to motivate artistic production. The program can be stated the other way around too: self-development leads to improved artistic production. Here I am trying to motivate self-development, though it too is an end in itself. I set for myself two ends.[8]

If one circumscribes as one's art work the production of art objects and one's self-development in a mutually enriching complex, then the objectivist rejection of subjective considerations as part and parcel of a work of art cannot be maintained.

While my personal program has an argumentative bearing on the objectivist account of "works of art" I have not come to formulate it in the first instance in order to criticize the objectivist view. Self-development, which concerns *how* one experiences the world rather than *what* one experiences of it, may be elucidated more concretely by recounting a personal episode, which was the impetus for the development of my personal program to begin with.

Though I have long been exposed to the visual arts, I was not particularly visually sensitive until 1971 at the age of 28. While visiting the studio of a friend and being surrounded by her large canvases, I experienced myself inhabiting the space visually depicted in them. It corresponded to and somehow forged an "inner" space with which I somehow felt familiar. It was a space in which I was to journey. I suddenly became visually much more highly sensitive, and, as a consequence, needed to paint — to pursue the journey — and became a painter. Since that time I have experienced the world differently. This experience was not unlike that described by Abraham Maslow as a peak experience.[9]

The question that immediately arises is, How does one judge self-development to be progressive? One can judge self-development (as opposed to mere change) if one comes to experience more clearly, more expansively, that is, more perspicuously.[10] I regard perspicuity as that felt quality of experience connected with the recognition that at a prior stage of development one was in a kind of experiential cocoon, a constricting envelope, perhaps, too, an incubator.

Artists who "freeze" their artistic development by replicating an art object repeatedly, perhaps because its prototype was well received by the public, "freeze" their self-development, just as "freezing" a moment in a human relationship in order to possess it indefinitely will destroy the relationship. Self-development exists in process, and the unfolding of *how* we experience may be thwarted. Consider the following personal episode by way of illustrating how one might "freeze" one's self-development.

In the spring of 1964 I was walking alone at the foot of Mount Vesuvius and unexpectedly came across a grouping of bushes, greenery, and flowers in which were a sheep and several lambs. It was a most beautiful sight. I was overcome by a religious feeling. I experienced a connectedness with the environment that was both exhilarating and frightening. Given my intellectual commitments at the time (I was a student of analytic philosophy), I was unable to make sense of this kind of experience. I cut the experience short, choosing not to be present for it. My rational faculties were summoned, and questions began to be put, doubts felt; the moment passed. Later I even chastized myself for permitting myself to experience such a "silly" thing.

I have come to think of this episode in terms of resisting not the objects of the environment, but a change in my sensibilities, in my *way* of experiencing. I resisted this change out of fear. The experience seemed to promise a loss of self-control, an entry into unfamiliar psychic and perhaps other territory.

In the visual arts the convention of restricting the notion of work of art to paintings, drawings, sculptures and the like was altered by the American Action Painters who considered their resultant physical items as part of "act-paintings" whose metaphysical substance was regarded as identical to the experience of life. Harold Rosenberg puts it this way:

> At a certain moment the canvas began to appear to one American painter after another as an arena in which to act — rather than a space in which to reproduce, re-design, analyze, or "express" an object, actual or imagined . . . the big moment came when it was decided to paint . . . just to PAINT . . . The painting itself is a "moment" in the adulterated mixture of (the artist's) life . . . *The act-painting is of the same metaphysical substance as the artist's existence.* The new painting has broken down every distinction between art and life.[11]

The idea of circumscribing one's acts as part of one's works of art is closely related to what John Dewey describes as having *an* experience. He says:

We have *an* experience . . . when the material experienced runs its course to fulfilment. Then and only then is it integrated within and demarcated in the general stream of experience from other experiences. A piece of work is finished in a way that is satisfactory; a problem receives its solution; a game is played through; a situation, whether that of eating a meal, playing a game of chess, carrying on a conversation, writing a book, or taking part in a political campaign, is so rounded out that its close is a consummation and not a cessation. Such an experience is a whole and carries with it its own individualizing quality and self-sufficiency. It is *an* experience.[12]

One may circumscribe the process of making, the resulting objects and the effects caused by the objects as "the artwork." In this way one may circumscribe self-development as a potential creative activity.

IV

It has often been pointed out that while one is making an art product one characteristically encounters unpredicted emergent features. Whether such features are simply unintended but predictable in principle (in accord with some physical or psychological regularities), or whether such features are unpredictable in principle depends, of course, upon whether there exist deterministic laws of the appropriate kind. But the phenomenon of emergent unintended features is well known. Sometimes emergent features are welcome, sometimes not.

Wassily Kandinsky was converted to non-figurative painting when, upon accidently leaving a figurative painting upside down in his studio, he returned only to see it in terms of an arrangement of lines, shapes, colors, that is, its formal aesthetic properties. Such was the coming of abstract art.

Picasso tells a story that illuminates emergent but unwelcome features in connection with a cubist painting being made by Braque. He says,

> I remember one evening I arrived at Braque's studio. He was working on a large oval still life with a package of tobacco, a pipe, and all the usual paraphernalia of Cubism. I looked at it, drew back and said, "My poor friend, this is dreadful. I see a squirrel in your canvas." Braque said, "That's not possible." I said, "Yes, I know, it's a paranoiac vision, but it so happens that I see a squirrel." . . . Braque stepped back a few feet and looked carefully and sure enough, he too saw a squirrel . . . Day after day Braque fought that squirrel. He changed the structure, the light, the composition, but the squirrel always came back . . . However different the forms became, the squirrel somehow managed to return. Finally, after eight or ten days, Braque was able to turn the trick and the canvas again became a package of tobacco, a pipe, a deck of cards . . . .[13]

Just as a single work may give rise to emergent features, so too may a *series* of related works, and these may become apparent to an artist only when the series is viewed as a whole. These features may challenge, and to some measure control, not only the development of an artist's autonomous work but also aspects of the artist's self-development in related domains of life.

If there is an awareness and willingness on the part of the artist to recognize emergent features and to identify with them, one may conceive of one's own personal *becoming* partly in terms of the *becoming* of one's art objects. I regard the making of art objects as part of this larger process of becoming, one that is continuous and open-ended. That, more fully, is the art work.

Sometimes artists and their making of art objects become suffused to the degree that artists become extentions of their work. John Dewey characterizes such moments as those when there is "complete interpenetration of self and the world of objects and events. Instead of signifying surrender to caprice and disorder, it affords our sole demonstration of a stability that is not stagnation but is rhythmic and developing."[14] It is this periodic At-Oneness (as I shall call it) and subsequent detachment of artist from art object, together with the emergent features of the art object, that can provide direction for the artist's work and his life.

The notion of At-Oneness is difficult to articulate, owing, I think, to our inherited vocabulary of subjectivity and objectivity.

Dewey tries to do it in terms of the interpenetration of the subjective and the objective. Chang Chung-yuan tries to articulate it
when he asks, "What is Tao-painting?" He answers,

> As we know, *Tao* is the ontological experience by which sub-
> jective and objective reality are fused into one. . . . When the
> Chinese artist says that he enters the spiritual court he speaks
> of the ontological experience, the state of no-thought. This
> experience leads inevitably to the *interfusion of subjective
> and objective reality*. This interfusion initiates the process of
> creativity, which in turn establishes unity in multiplicity, the
> changeless in the ever-changing. The artist who has reached
> this state of oneness is supported by all the powers inherent
> in multiplicities and changes, and his work will be far beyond
> what his egoform self could accomplish. Robert Henri
> (1865–1929), speaking of modern art, expresses somewhat
> the same idea: The object, which is back of every true work
> of art, is the attainment of a state of being, a state of high
> functioning, a more than ordinary moment of existence. In
> such moments activity is inevitable, and whether this activity
> is with brush, pen, chisel, or tongue, its result is but a by-
> product of this state, a trace, the footprint of the state. The
> work of art is, indeed, the by-product of a state of high func-
> tioning. This state of spiritual exaltation is fundamental to
> creative activity, while skills and measurements are secondary.
> It is the manifestation of an ontological experience.[15]

Favorable conditions for "At-Oneness" may be fostered. Ironically, directly trying to possess it can be an obstacle to experiencing
it, because the possessive impulse affirms the place of the "self" in
opposition to the "other."[16]

The occasional At-Oneness between self and other, between
artist and art-product, is episodic in an unfolding history of self-
development and artistic development. Despite the rarity of such
states, however, it should be noted that their existence calls into
question a fixed and sustaining distinction between self-development and artistic development.

Thus, the first formulation of my personal program should be
regarded as approximate, and requiring further amendment. I said

that my painting is instrumental to my self-development and my self-development is instrumental to my painting, suggesting thereby that these realms, while related, are always distinct. Sometimes, though, they are not so distinct.

In a recent comment in the journal *Leonardo* John Albin Broyer has remarked that

> There appears to be a deep religious undercurrent flowing through Krausz's [account], and especially in his frequent use of the term "at-one." It would seem that the obvious correlation between his use of "at-one" and the religious use of the term *atone*ment to describe people's quest for overcoming fragmentation in their lives through reconciliation, is more than coincidental. George Santayana and John Dewey are among the few contemporary philosophers with sufficient intellectual rigor and aesthetic sensitivity to develop viable hypotheses about the relationship of aesthetic consummation and religious meaning. Krausz could suggest the possibility that creative self-development through artistic-aesthetic expression may be one avenue for understanding the "religious" experience of atonement. If so, the import of art for self-development would make the "beauty" of art a functional *consequent* of its human meaning, rather than its *antecedent*, as is often supposed. "Beauty" would reside neither in the art object nor in the beholder alone, but would emerge as a definition of individual (and in "great" art, social) atonement.[17]

I find Broyer's observations most illuminating, and there may indeed be an affinity between At-Oneness and religious experience. Yet I am not sure whether religious vocabulary is the only or the best vocabulary in terms of which to describe At-Oneness. It has been variously described by such authors as George Santayana, John Dewey, Aldous Huxley, Anaïs Nin, Martin Buber, and Abraham Maslow. For my purposes it is sufficient to indicate that it involves a suffusion of the subjective and the objective, an interpenetration of one with the other, as Dewey puts it. That, surely, is enough to cause problems for the objectivist who wishes to maintain a strict separation between the objective and the subjective.

Now, even if one were to reject Dewey's characterization of the interpenetration of the subjective and objective, even if one were to go so far as to deny the intelligibility of what Dewey is trying to characterize, it remains that one might circumscribe both artistic production and self-development as one's art work, as have the American Action Painters and Duchamp and his conceptualist followers. So, circumscribing the parameters of "work of art" is not a matter of arbitrarily bracketing random objects of experience. It involves a concerted fashioning of experience in much the way Dewey describes our making of experience *an* experience.

David Carrier has made a pointed observation about the creation of art works by a Deweyan circumscription. He says, "If one accepts Krausz's account, does it follow that a painting is not identical with a physical object, the pigment on canvas? And can spectators in a museum who see a painting made by Pollock be said to fully appreciate the artwork made by that artist?"[18]

Here one must be clear about the object of evaluation. A resultant physical painting, initially produced within the context of an act-painting, may be evaluated in its own aesthetic terms, just as a building initially conceived as part of a larger urban design may be evaluated in its own aesthetic terms. By suggesting that a painting may be initially conceived as part of a larger work of art I do not rule out the aesthetic appreciation and evaluation of that painting on its own.

The question whether one might appreciate and evaluate the original act-painting in its full intensional context is made more difficult, obviously, by a relative inaccessibility of that full intensional context. But something of a rational reconstruction can be offered here, it seems to me, based on pertinent historical information — the sort of information, for example, that permits us to characterize the American Action Painters and the conceptualists as trying to do something quite untraditional to begin with.

Carrier's query is closely related to one of James Munz. He has noted that:

> [Krausz] appears to assume that the standards appropriate for judging one's artistic production are also appropriate for judging one's self-development. But, it would be no more or less appropriate to judge one's artistic production in terms

that satisfy one's standards of self-development. If artistic production and self-development are separate ends, there is no reason that their standards are identical.[19]

I agree with Munz that there is no logical necessity, so to speak, in the identity of standards for artistic development and self-development. That each, at the least, is instrumental for the other may be felicitous. Felicity would be enough. But two further considerations suggest more than felicity. (1) If At-Oneness involves an interpenetration of the subjective and the objective, the sustained distinction between artistic development and self-development — and thereby their putatively independent standards — is called into question. And (2) to recall Broyer's interesting reference to atonement, there may well be an overriding aim to the standards of artistic production and self-development, namely, the quest for overcoming fragmentation in life through reconciliation. The standards for such an aim would have to be spelled out, to be sure, but they might well be of a kind that would loosely overarch the standards for artistic production and self-development. In short, the introduction of an overriding aim would make the standards for artistic production and self-development non-felicitous. But again, felicity would be enough.

Should there be no felicity, though, the dual pursuit of artistic production and self-development would have to be modified; this might be done by pursuing artistic development only, by pursuing self-development only, by pursuing reconciliation through other means, or none of these. It has been my contention, though, that the first two of these aims are integrally connected.

I have already rehearsed Briskman's logical reasons for conceptualizing self-development as an *un*intended consequence or as a *by*-product of creative activity. He also has a psychological reason for doing so. That is, if the self-developmental consequences of creative activity are the objects of consciousness, self-development itself would be impaired, much, perhaps, as the centipede's walking might be impaired if it worried what a particular leg might be doing. Here Briskman confuses having an overriding aim with having that aim as a continual object of consciousness. I do not suggest that while engaging in creative activity one continuously focus on self-development, nor, for that matter, on artistic development. Once we see, though, that there is a structural relation between artistic production and self-development, it is a matter of personal preference how much one fruitfully dwells on self-development or artistic production as an object of consciousness while one is creating.

It should be noted that there is a certain congruence between Briskman's object-centered view and mine. I have argued, following the lead of Duchamp and the others, that we may circumscribe as creative products those processes which we usually are not inclined to regard as products, namely the process of the making, the process of self-development. Such life processes are often thought to precede the production of creative objects. I have suggested that they may constitute creative objects. So, I believe that the real difference between Briskman and myself turns on the question of what constitutes a product, that is, on whether the processes that I allow as constituting products must precede what is legitimately circumscribed as the product. At this point I regard the onus to be on Briskman to make good his view that such processes must precede rather than constitute such products.

To summarize, what I make conditions how I experience, and how I experience conditions what I make; my more perspicuous ways of experiencing give rise to yet more perspicuous art objects. It is by virtue of such perspicuity that I speak of self-development rather than mere change. However, my work as a painter is not merely catalytic for my self-development, because the autonomous emergent features of my paintings, especially when I am At-One with them, populate and enlarge my inner world.

I have called upon this personal program to aid me in making more sense of my artistic activity. As Popper suggests, one develops and alters one's program as time passes, modifying both its aims and strategies to suit special concerns and circumstances. While continually dwelling upon a personal program may actually divert one from the aims it was intended to foster, the articulation of a personal program can help us see the variability of creative products and help locate ourselves in relation to those products.

Notes

*This essay is a version of one presented to the Eastern Division of the
Society for the Philosophy of Creativity, American Philosophical Associ-
ation, Washington, D.C., December, 1978. It is dedicated to Constance
Costigan and to the memory of my father, Laszlo Krausz (1903–1979).

1. E.H. Gombrich, *Art and Illusion* (Princeton: Princeton University Press,
   1969).
2. Karl R. Popper, *Objective Knowledge* (New York: Oxford University
   Press, 1975).
3. Karl R. Popper, *Unended Quest* (LaSalle: Open Court, 1976).
4. K.R. Popper and J.C. Eccles, *The Self and Its Brain* (New York: Springer,
   1977).
5. Ursula Meyer, *Conceptual Art* (New York: E.P. Dutton, 1972), p. ix.
6. Meyer, p. xv.
7. Popper and Eccles, op. cit., p. 91.
8. I am grateful to James Munz for this formulation. See *Leonardo* 13
   (1980): 350.
9. A.H. Maslow, *Religions, Values, and Peak Experiences* (New York: Viking
   Press, 1963).
10. J.N. Findlay, "The Perspicuous and the Poignant," in *Aesthetics*, Harold
    Osborne, ed. (London: Oxford University Press, 1972).
11. Harold Rosenberg, "The American Action Painters," *Art News* 51 (1952):
    22–23, 48, emphasis added.
12. John Dewey, *Art As Experience* (New York: Capricorn Books, 1934),
    p. 35.
13. F. Gilot and C. Lake, *Life With Picasso* (New York: McGraw-Hill, 1964),
    pp. 68–69.
14. John Dewey, op. cit., p. 19.
15. Chang Chung-yuan, *Creativity and Taoism: A Study of Chinese Philos-
    ophy, Art, and Poetry* (New York: Harper Colophon Books, 1970),
    pp. 203–204.
16. See E. Fromm, *To Have or To Be?* (New York: Harper and Row, 1976).
17. John Albin Broyer, Letter to the Editor, *Leonardo* 13 (1980): 350.
18. David Carrier, Letter to the Editor, *Leonardo* 13 (1980): 262.
19. James Munz, Letter to the Editor, *Leonardo* 13 (1980): 350.

ON THE DIALECTICAL PHENOMENOLOGY
OF CREATIVITY

ALBERT HOFSTADTER

I would like to present to you the idea of thinking about creativity in terms of the basic concepts of dialectical phenomenology. It is unnecessary to stop to examine the notion of dialectical phenomenology. Illustration of it is already before us in the shape of Hegel's *Phenomenology of Spirit* and his later writings about the philosophy of mind. We need not adopt Hegel's own monistic spiritualistic metaphysics in taking over the basic ideas of dialectical phenomenology, nor need we presuppose a logically necessary and single-track line of development, whether in individual life, social life, or world history. What I wish to suggest is only the dialectical structure and meaning of the creative process, with the thought that study of this process in terms of dialectical phenomenology could open it up to a genuine mode of comprehension.

It is generally agreed that creation is not a one-sided mechanical making according to a fixed formula or routine. So, for example, creative photography is not like the process by which the passport photographer takes your picture. He already knows what content and what form he wants the picture to have, and he needs only to fit you, as subject, into the frame to produce your picture. He has a content and form already decided beforehand, and has only to fit the new matter into it. He could take the same picture of anyone else; and the degree to which the process is routine is measured by the comic effect that would be produced if he substituted a horse or a camel for a human being. The creative photographer, on the other hand, finds that the very idea of a picture is already a challenge. He does not know beforehand what content it is to have, what its form is to be. His task, indeed, is precisely to find a content and a form and the unity of the two, so as to determine the being of a picture. "Picture" first gets a meaning for him in this process of finding/inventing the new unity of form, content, and subject. Instead of following a routine he has to develop the

D. Dutton and M. Krausz, eds., The Concept of Creativity in Science and Art, pp. 201-208.
© 1981 Martinus Nijhoff Publishers, The Hague/Boston/London. All rights reserved.

new totality itself: the new unity of form, content, and matter, which is to be the new "picture."

What is it that differentiates the routine from the creative photographer? The routine photographer, like the passport photographer, already has his general form-content unity fixed, and he swings his lens around to the prefixed type of subject so as to produce the normally expected result — the expertly done photograph, the arty picture. The creative photographer, to the contrary, is first in a quandary. He does not have a prefixed form-content idea with its correlated subject-matter class. He looks upon all that as not his dish of tea; he finds himself estranged from it, finds it alien to his spirit. He has to find in the whole situation a form-content unity and its correlative subject-matter to which he can give himself and which will give itself to him. He is in search of a new totality which will be own to him and to which he will give himself as its own. He is in search of a new reconciliation of the factors of opposition and difference — form, content, subject-matter on the one hand, and his own living spirit on the other — which he can experience as real truth, as genuine actuality, a totality in which inner and outer factors have come together in a living, not a mechanical, unity.

I use the example of the photographer because it is so easy to see the difference almost immediately, since the routine mechanical photographer depends chiefly on mechanical means — predetermined content, form, subject, instrument, equipment — all of them so ready at hand that anyone can be such a photographer, whereas the creative photographer is in the same general fix as any creative being: everything around him has fallen into flux, he himself swims in the midst of the turbulent stream without any shores, and he has to find his way to safety by bringing the flux into the rhythm of a living form.

But the photographer is only the image of us all. Wherever we are, in private life or public, art or religion, action or thinking, the same general problem faces us. We can approach the context with our prefixed formula for the given sphere of life — prefixed meanings, forms of procedure, subjects to which they are applied, habits of feeling, willing, and thinking, capacities and skills — and try to put things together in the same way as before. And as long as this works, we can persist in it. When, however, life inside or out rebels against this mechanical way of constraining it, we are confronted with the problematics of creativity. We have to become creative, since we have to find and invent a new totality of form-content, matter, and spirit which will enliven the life process.

This life-quandary is widespread today. The people who are most involved in it, that is, the younger people who are now confronted with the task of determining the form and content of their life, experience the rebellion of life within their own selves, and they find themselves in profound alienation from much of the traditional culture. So they are in the midst of the crisis that is characteristic of creativity generally: the crisis of estrangement, alienation, otherness, difference.

This language of alienation, estrangement, otherness, and difference brings us directly into touch with the context of dialectical phenomenology. If anything is characteristic of the dialectic it is the appearance of the phenomenon of alienation or estrangement, or more generally of difference, opposition, and otherness. These are one half of the story of the dialectic; the other half is given in the correlative names for unity: reconciliation, appropriation, and ownness. Hegel says "Versöhnung," Heidegger "Ereignis," and both ultimately come to the same thing as "appropriation" and "ownness." It is ultimately in the structures of ownness that human creativity shows itself and its products. The meaning of creativity lies in the attainment of genuine ownness.

In thinking about creativity we tend to think first of the artist as creator and his activity as creation. In earlier days one thought first, rather, of God and of His act of creation. Our present tendency is a reverberation of the romantic movement in life and art which placed the artistic individual at the center and which put special accent on his subjectivity as the source of truth rather than on submitting his subjectivity to a reality outside him as foundation of truth. But creativity belongs everywhere, and perhaps it would be better if we looked more frequently to spheres outside art for subjects of investigation. We shall recollect that Whitehead thought creativity to be so pervasive, present in every pulse of process that gave rise to a new actual entity, that he declared it to be the first and most universal of all the categories. Therefore, in what immediately follows, I would like to bring to your attention as an instance of creativity a nonartistic matter, specifically the apparently unlikely matter of the master-slave relationship. Since it is one of the most achieved parts of Hegel's phenomenology and has been much looked at in recent years, and since it lies at the basis of the possibility of advanced socialized life, it is of particular interest as a means of directing our attention to creativity in life and experience apart from art.

The master-slave or lord-bondsman relationship arises out of the struggle for recognition. Two egos stand in confrontation. Each, as an ego, seeks recognition by the other as being "I." Each seeks to be the essence of their mutual relationship. Each needs an other to be its own: the one needs to appropriate the other, needs the other to be reconciled within a unity in which the ego is recognized as the essence. In an other alone can the ego find its own, so that it can truly be ego. Its basic drive as ego is to be with its other as its own. The problem is: how is this being-with to be achieved?

The two egos are not themselves self-conscious about this need they share. It is we, who look on from "above the battle" (for we have already been through it and are now reflecting backwards), who can apprehend with conscious awareness what is going on there. The two involved egos are there in the throes of their own desires and passions, basically the passion to be free — that is, to find selfhood in and with what is not self, and in this case particularly to find self-recognition in the other. They do not yet know this passion to be theirs — they are, rather, inside it, moving in and moved by it; the knowledge of it will come, and can only come, as it gets realized, so that they can see in the realized form what it is they were struggling for. (As Croce maintained: the creative one does not know what he has been striving to create until he has created it.)

Before the master-slave relationship developed in human culture there was the immediate relationship of being-with, as in the hunting band, where no one is chief, ruler, lord, but everyone has his say and his part in determining what is to be done. The life of the Mbuti pygmies in the African Congo region (as described by Colin Turnbull in *Wayward Servants*, Garden City, New York: Natural History Press, 1965) gives a beautiful picture of what a life is like which dispenses with the master-slave foundation — although, perhaps, a closer look would indicate how powerfully the forest itself is given the role of master in relation to all members as its servants. There is here a necessary suppression of individuality so as to prevent any particular member from asserting himself — no matter how great his talent, skill, even genius — beyond the average level of sociality. This book is indeed a fascinating story of how a people can create, with genuine and continuing creativity, a form of social life that is essentially free, equal, and fraternal within itself, but also, within the confines of a very limited geography and socio-psychological topography. But all advanced culture depends on the prior establishment of political power and the relation of ruler to ruled. Much of human history has lain in the production and transformation of this relationship towards the point at which eventually its asymmetry can be entirely removed.

The theme I wish to stress, then, is that the master-slave relationship, as such, is something that is created in the struggle between the confronting egos as the mode of unity — the mutual appropriation, reconciliation, or ownness — which sublates (*aufhebt* — cancels, preserves, while transcending) the opposition of the two agents, raising them up into being the subjects of a new totality, with a new content and form, which gives (albeit only a passing one, nevertheless) a fulfillment to their mutual need for recognition.

The idea of master-slave, lord-bondsman, was an indispensable creation in the history of the development of human society, and it still remains as an indispensable creation in the history of human individuality. Much of childhood's suffering, as well as its fulfillment and security, lies in the recognition it is compelled, by its own weakness, to give to the adult-parent ego, finding its own selfhood in the maturity, strength, and above all credible authority of the parent. The suffering is greater when that authority later comes into question and when, therefore, a new creative art of social appropriation of self and other is needed — i.e., the achievement of a new unified totality of content, form, and the subjectivity of its own selfhood. This is part of the search for one's identity as outcome of the identity crisis and is the clear counterpart, in the actual struggle for life, of the artist's search for artistic identity as he grows out of dependence on earlier forms, traditions, and masters and finds himself cast into the outsideness and indeterminateness of new existence.

Let us try to see how the master-slave relationship — despite the negative attitude we presently have towards the subjection of ruled to ruler, political subject to despotic king or dictator, serf to lord, slave to slaveowner — was nevertheless an actual creation, a step forward into novelty and constructive order.

The egos, to begin with, are isolated within themselves; each, outside the other, is not the other, and the other is not it; they are really different. But difference among egos cannot remain as mere otherness: the different is the strange and alien, and what is strange and alien is, or rapidly becomes, hostile, an enemy. The further off people are from us, the easier it is for us to regard them, and for them to regard us, as actually or potentially hostile. We approach one another with caution, for, being distant, without nearness, we are not neighbors, and so it does not grieve one when the other is mishandled; on the contrary, even we, who have been educated and disciplined in the values of the high religions of the East and the West, find it easy enough to take the position of master to the other as slave, determiner to determined.

The two thus become enemies. It is tempting to think that they could have started at the very beginning as friends; but one does not know how to think "friend" except through thinking "enemy." Friendship needs the long discipline of living together in the security of a social environment that cares for both. Enmity needs only the initial encounter of those who are other to one another.

The two become enemies, however, because they need one another. For outside the relationship of ego to ego, self to self, the ego cannot find the ownness it needs in order to be itself. Without the alter ego the ego is outside the *Ereignis* — outside the appropriation of being to being, outside the reconciliation, the *Versöhnung,* which it needs in order to come to its own. The step into enmity is already the first step on the road to the creation of a genuinely human life. Instead of passing one another by, the two are attracted to each other and in need of one another. Egos are beings who, being an "I," can have a "mine." And this means that a further possibility — amazingly new, fecund, fraught with the deepest potentialities of existence and truth — begins to appear. For, if an ego can apprehend and appropriate something as "mine," it is on the point of being able to turn around and *give* itself also as "thine." It needs only another ego, in whom it can recognize also the power of owning, to which it is able to yield.

Both the "mine" and the "thine" are necessary for the new unity that comes about between separate egos who have become enemies. In its primitiveness each can think first only of possessing the other as its own, as "mine." To possess an other ego as "mine" means that the other ego must *give* itself to me, so that I am its "thine." Neither has as yet learned how to give itself. They know only how to demand and try to take what is other as their own. Each wants the other to belong to it, but in the way in which an ego can belong to an ego — that is, by recognition, acknowledgment, giving of itself to the other (as gift) so that the other's self is its own self, its own essence — what is mine is thine, even to the point of my own self. This is what the one ego wants of the other; and the struggle to the death to achieve it is itself the creative process of realizing this new, profound, soul-shaking idea.

Both egos are engaged in this process. The creative agent is neither the one alone nor the other, but the two of them in a mutual structure of consciousness of other and of self. There is a We in process of self-formation, creating itself in and out of the I's, first by developing in and through their animosity as alien and estranged, and then by the struggle of this animosity to bring itself into being in the definite shape of the master-slave condition.

Underlying the process is the fundamental need and desire, the passion to be free, to find ownness and otherness, self-recognition in the other, or, as Heidegger's phrase goes, to be "gathered in the appropriation." This passion is the source of the dialectical structure of the creative process. Gathering in appropriation needs, first of all, the *Enteignung*, the de-appropriating which is at the same time a readying of the appropriate to the appropriate, so that there can be a true appropriation, an *Ereignung*, in which the others find their own. From the mere sameness, the abstract difference, of the two egos, there had to develop first their real difference, their opposition and alienation, precisely in order that the struggle should begin in which they might find in each other their own.

I have not the space to specify in detail how not only the master finds his own in the slave, but also how the slave first begins to

208

find what is truly *man's* own, and not merely this individual's own, in the discipline of rule, so that a creative process is started that looks towards reconciliation at a level of ultimacy.

The struggle of the artist with his medium, and also with his society and culture, his artistic competitors, his own past education, his habits and the pregiven cast of his mind, the temptations of popular success, and so many other factors of difference, is also a struggle to be with all these as own, to transform others and self into a new shape, gathering them all into a new appropriation in which they fit so as to be able to live in freedom.

The struggle of every creative person — and this means every person so long as he or she tries to continue to live in a meaningful way — is this same dialectic of plunging into the alienation in order to reach towards the appropriation, opening one's self and one's situation to the factors of difference, opposition, and estrangement so as to permit the possibility of a new living gathering.

As the bridge gathers the opposing shores of the river, the country and city, the fields and streets, the one life and the other, into a reciprocal totality of ever-moving life, so the creative person bridges otherness to gather what is alien into a new ownness.

The categories we must use to comprehend creativeness, wherever it is, are those of identity, differentiation, and recovery, of estrangement and reconciliation, the alien and the own, the gathering of differents and opposites into the appropriation of a reciprocal ownness.

These are the categories of dialectical phenomenology, which have yet to be tried for their power of enlightenment.

University of California
Santa Cruz

NAME INDEX

Abrams, M.H., 73n.
Agassi, Joseph, 127n., 152n., 154n.
Alloway, Lawrence, 127n.
Aristotle, 48, 52, 63, 73n., 120

Bach, C.P.E., 120
Bach, J.C., 120
Bach, J.S., 8, 83, 120, 127n., 188
Bachelard, Gaston, 185n.
Bachrach, A.J., 107n.
Bacon, Francis (painter), 113, 127n.
Bacon, Francis (philosopher), 152n.
Barron, F., 151n.
Barthes, Roland, 166
Basshō, 59
Beardsley, Monroe, C., 70, 73n., 187
Beethoven, Ludwig van, 76, 81, 137, 188
Bergman, Ingmar, 115
Bergson, Henri, 175, 185n.
Berkeley, George, 113, 120
Bertalanffy, Ludwig von, 185n.
Beveridge, William I., 185 n.
Bower, G.H., 185n.
Braque, Georges, 15, 193-94
Brecht, Bertolt, 116
Briskman, Larry, 187-89, 198-99
Broadbent, Donald I., 185n.
Broyer, John Albin, 196, 198, 200n.
Buber, Martin, 196
Bunzel, Ruth L., 73n.

Campbell, Donald, 155n.
Carrier, David, 197, 200n.
Carroll, Donald, 73n.
Cassirer, Ernst, 160, 184, 185n.
Cézanne, Paul, 12, 15
Chandler, Raymond, 116
Chang Chung-yuan, 195, 200n.
Chaplin, Charlie, 115-16, 127n.
Chomsky, Noam, 160, 166, 185n.

Cocteau, Jean, 127n.
Coleridge, S.T., 153-54n.
Collingwood, Robin G., 54, 64-65, 73n.,
    153n.
Copernicus, Nicolas, 92-3, 97, 102, 105,
    120, 175
Coulomb, Charles-Augustin de, 33
Croce, Benedetto, 204

Dali, Salvador, 127n.
Dalton, John, 154n.
Darwin, Charles, 23-26, 42, 71, 147, 150,
    155n., 175
DeMille, Cecil B., 115
DeSantillana, G., 17n.
Descartes, René, 29, 102, 152n.
Dewey, John, 160, 192, 194-97, 200n.
Dingle, H., 107n.
Dirac, Paul, 16, 17n.
Donne, John, 7
Duchamp, Marcel, 189-90, 197, 199
Dufrenne, Mikel, 73n.
Duhem, Pierre, 92
Dukas, Helen, 154n.
Dürer, Albrecht, 15

Eccles, J.C., 200n.
Einstein, Albert, 2, 13, 76, 95-99, 101, 105,
    107n., 109, 113-14, 118, 122, 127n.,
    128n., 137, 141, 143, 146-47, 152n.,
    154n., 162, 164-65
El Greco, 83
Eliot, T.S., 62, 72n., 125, 128n.
Eliot, Valerie, 128n.
Elliot, Charles H., 185n.
Euclid, 164, 180
Eveling, Stanley, 154n.

Fairbanks, Douglas, 116
Faraday, Michael, 30

210

Faulkner, William, 116
Findlay, J.N., 200n.
Fischer, Bobby, 113
Fitzgerald, F. Scott, 116
Ford, Edmund Brisco, 25
Ford, John, 115
Foucault, Michel, 166
Freud, Sigmund, 8, 14, 62, 113, 121
Frey, Gerhard, 185n.
Fromm, Erich, 200n.
Fuller, R. Buckminster, 15

Galilei, Galileo, 92-93, 102, 143
Gallie, W.B., 189
Gauss, Carl Friedrich, 130
Geach, Peter, 29
Getzel, J.W., 126n.
Ghiselin, B., 153n.
Gibson, J.J., 46n.
Gilbert, William, 30
Gilot, Françoise, 200n.
Giotto, 76, 78, 81
Goffman, Erving, 21, 26-27, 39, 45n.
Golovin, N.E., 134, 151n., 152n.
Gombrich, Ernest H., 17n., 127n., 153n.,
    155n., 187, 200n.
Goodman, Nelson, 189
Goodrich, Lloyd, 128n.
Graves, Robert, 62, 72n.
Greenberg, Joel, 127n.
Griffith, D.W., 115-16
Grünbaum, Adolf, 96, 107n.
Guilford, J.P., 110, 126n., 134, 151n

Hadamard, Jacques, 13, 17n., 151n.
Hardwicke, Cedric, 116
Harré, Rom, 45n.
Hart, H.L.A., 104, 108n.
Harvey, William, 12, 29, 33
Hebb, Donald O., 185n.
Hecht, Ben, 116
Hefferline, R.F., 107n.
Hegel, F.W., 201, 203
Heidegger, Martin, 203, 207
Helmholtz, Hermann, 129
Helmont, Johannes Baptista van, 29
Henri, Robert, 195
Hertzmann, Erich, 128n.
Higham, Charles, 127n.
Hilgard, Ernest R., 185n.

Hoffman, Banesh, 154n., 155n.
Hogarth, William, 6
Hopper, Edward, 128n.
Hospers, John, 73n.
Houseman, John, 116
Housman, A.E., 58
Howard, Leslie, 116
Hudson, William, 152n.
Hume, David, 114
Husserl, Edmund, 99
Huxley, Aldous, 116, 196

Isherwood, Christopher, 116

Jackson, P.W., 126n.
James, Brennig, 17n.
James, William, 78, 73n., 94, 100
Jarvie, I.C., 127n., 128n.
Jung, Carl Gustav, 17

Kael, Pauline, 127n.
Kandinsky, Wassily, 193
Kant, Immanuel, 28, 84, 178
Keaton, Buster 115
Kekulé, August, 29
Kepler, Johannes, 21, 24, 29, 92-93, 99,
    102, 143
Keynes, John Maynard, 164
Kluckhohn, Clyde, 166
Koestler, Arthur, 17n., 21, 45n., 62, 70,
    72n., 102, 107n., 121, 126n., 155n.,
    159
Kottenhoff, H., 107n.
Koyré, Alexandre, 166
Krausz, Michael, 196-97
Krebs, H.A., 151n.
Kris, Ernst, 15, 17n.
Kuhn, Thomas S., 17n., 165-66, 188
Kuniyoshi, Yasuo, 153n.

Lakatos, Imre, 151n., 152n., 153n.
Lake, C., 200n.
Langer, Susan K., 185n.
Lang, Paul Henry, 128n.
le Corbusier, 15
Leibniz, Godfried Wilhelm, 99
Leigh, Vivien, 116
Leonardo da Vinci, 15
Lévi-Strauss, Claude 27, 46n., 166
Linnaeus, 175

Locke, John, 29
Louis XV, 4, 5
Lucie-Smith, Edward, 73n.
Luria, A.R., 160, 185n.

Mach, Ernst, 105
Mann, Thomas, 116
Maslow, Abraham H., 191, 196, 200n.
Maxwell, James Clerk, 35, 143
May, Rollo, 75, 85, 89n.
Mendelssohn, Felix, 127n.
Merleau-Ponty, Maurice, 185n.
Meyer, Ursula, 189, 200n.
Michelson, Albert Abraham, 96
Mill, John Stuart, 24
Moles, A.A., 73n.
Monod, Jacques, 132, 151n.
Morgenstern, O., 175
Morley, Edward W., 96
Morris, Desmond, 130, 151n.
Morris, Robert, 189
Mowrer, Orval H., 185n.
Mozart, Wolfgang Amadeus, 7, 84, 114, 123-24, 128n., 129, 132
Munz, James, 197-98, 200n.
Murphy, G., 185n.
Musgrave, A., 151n.

Newton, Isaac, 33, 86, 92-93, 97, 102, 105, 120, 122, 127n., 143, 162, 164-65, 175
Nilsson, S., 155n.
Nin, Anaïs, 196
Norman, Robert, 30
Novitz, David, 185n.

Oersted, Hans Christian, 2
Olivier, Laurence, 116
Osborne, Harold, 200n.

Pallach, Jan, 182
Pavlov, Ivan Petrovich, 158, 160, 163, 171, 173, 185n.
Peursen, C.A. van, 185n.
Piaget, Jean, 8-9
Picasso, Pablo, 83, 130, 193, 151n., 200n.
Pickford, Mary, 116

Plato, 9, 69, 73n., 97, 120, 154n.
Poe, Edgar Allan, 61, 72n.
Poincaré, Henri, 86, 92
Polanyi, Michael, 108n., 158
Pollock, Jackson, 197
Popper, Karl, 17n., 21-2, 26, 43-45, 45n., 46n., 91, 121, 127n., 128n., 133, 141, 151n., 152n., 154n., 155n., 164, 185n., 187-88, 190, 199, 200n.
Pound, Ezra, 125, 128n.
Pronko, N.H., 107n.
Ptolemy, 92, 102, 120, 175
Pythagoras, 2

Quine, W.O., 153n.

Rawlinson, James, 6
Razik, T.A., 134, 152n.
Rembrandt, 114, 127n.
Ricoeur, Paul, 174, 185n.
Rosenberg, Harold, 192, 200n.
Russell, Bertrand, 123-25, 128n., 142

Santayana, George, 196
Sartre, Jean-Paul, 133, 151n.
Schapp, W., 185n.
Schilpp, P.A., 107n., 108n., 154n.
Schumann, Robert, 127n.
Shahn, Ben, 139, 153n.
Shakespeare, William, 76
Shapiro, R.J., 152n.
Searle, John, 185n.
Sennett, Mack, 127n.
Seurat, Georges, 16
Shelly, J.H., 151n.
Skinner, B.F., 159-60, 185n.
Smith, C. Aubrey, 116
Snyder, F.W., 107n.
Sparshott, F.E., 72n.
Spinoza, Baruch, 99
Staats, Arthur A., 185n.
Staats, Carolyn K., 185n.
Storr, Anthony, 113, 122, 126n., 127., 128n.
Stratton, G.M., 95, 98
Stravinsky, Igor, 116
Stroheim, Eric von, 115

Taylor, C.W., 151n.
Tchaikovsky, Peter, 129
Thorpe, William H., 185n.
Tiselius, A., 155n.
Tolstoy, Leo, 40
Tomas, Vincent, 65, 70, 73n., 151n., 153n., 155n.
Toulmin, Steven, 40, 44-45, 46n., 153n.
Toynbee, A.N., 17
Turnbull, Colin, 204

Vaihinger, Hans, 168
Valéry, Paul, 61, 72n.
Velikovsky, Immanuel, 29
Vernon, P.E., 126n. 151n., 152n.
Vidor, King, 115

Vivas, Eliseo, 155n.

Wallace, Alfred Russel, 23, 24-26, 42, 71
Wallas, Graham, 128n.
Walpole, Hugh, 116
Welles, Orson, 115-16
West, Nathanael, 116
Whewell, William, 24, 45n.
Whitehead, Alfred North, 175
Wilder, Thornton, 116
Wimsatt W.K., 187
Wittgenstein, Ludwig, 109
Woodworth, R.S., 14, 151n.

Yeats, William Butler, 11
Young, Edward, 73n.